Narrative of a second expedition to the shores of the Polar Sea, in the years 1825, 1826, and 1827. Including an account of the progress of a detachment to the Eastward, by J. Richardson. (Appendix I-VII. Topographical and geological notices.)

John Franklin

Narrative of a second expedition to the shores of the Polar Sea, in the years 1825, 1826, and 1827. ... Including an account of the progress of a detachment to the Eastward, by J. Richardson. (Appendix I-VII. Topographical and geological notices; meteorological tables; observations on Solar radiation; by J. Richardson. Observations on the velocity of sound at different temperatures; by E. N. Kendall, etc.)

Franklin, John
British Library, Historical Print Editions
British Library
1828.
8º.

The BiblioLife Network

This project was made possible in part by the BiblioLife Network (BLN), a project aimed at addressing some of the huge challenges facing book preservationists around the world. The BLN includes libraries, library networks, archives, subject matter experts, online communities and library service providers. We believe every book ever published should be available as a high-quality print reproduction; printed on- demand anywhere in the world. This insures the ongoing accessibility of the content and helps generate sustainable revenue for the libraries and organizations that work to preserve these important materials.

The following book is in the "public domain" and represents an authentic reproduction of the text as printed by the original publisher. While we have attempted to accurately maintain the integrity of the original work, there are sometimes problems with the original book or micro-film from which the books were digitized. This can result in minor errors in reproduction. Possible imperfections include missing and blurred pages, poor pictures, markings and other reproduction issues beyond our control. Because this work is culturally important, we have made it available as part of our commitment to protecting, preserving, and promoting the world's literature.

GUIDE TO FOLD-OUTS, MAPS and OVERSIZED IMAGES

In an online database, page images do not need to conform to the size restrictions found in a printed book. When converting these images back into a printed bound book, the page sizes are standardized in ways that maintain the detail of the original. For large images, such as fold-out maps, the original page image is split into two or more pages.

Guidelines used to determine the split of oversize pages:

- Some images are split vertically; large images require vertical and horizontal splits.
- For horizontal splits, the content is split left to right.
- For vertical splits, the content is split from top to bottom.
- For both vertical and horizontal splits, the image is processed from top left to bottom right.

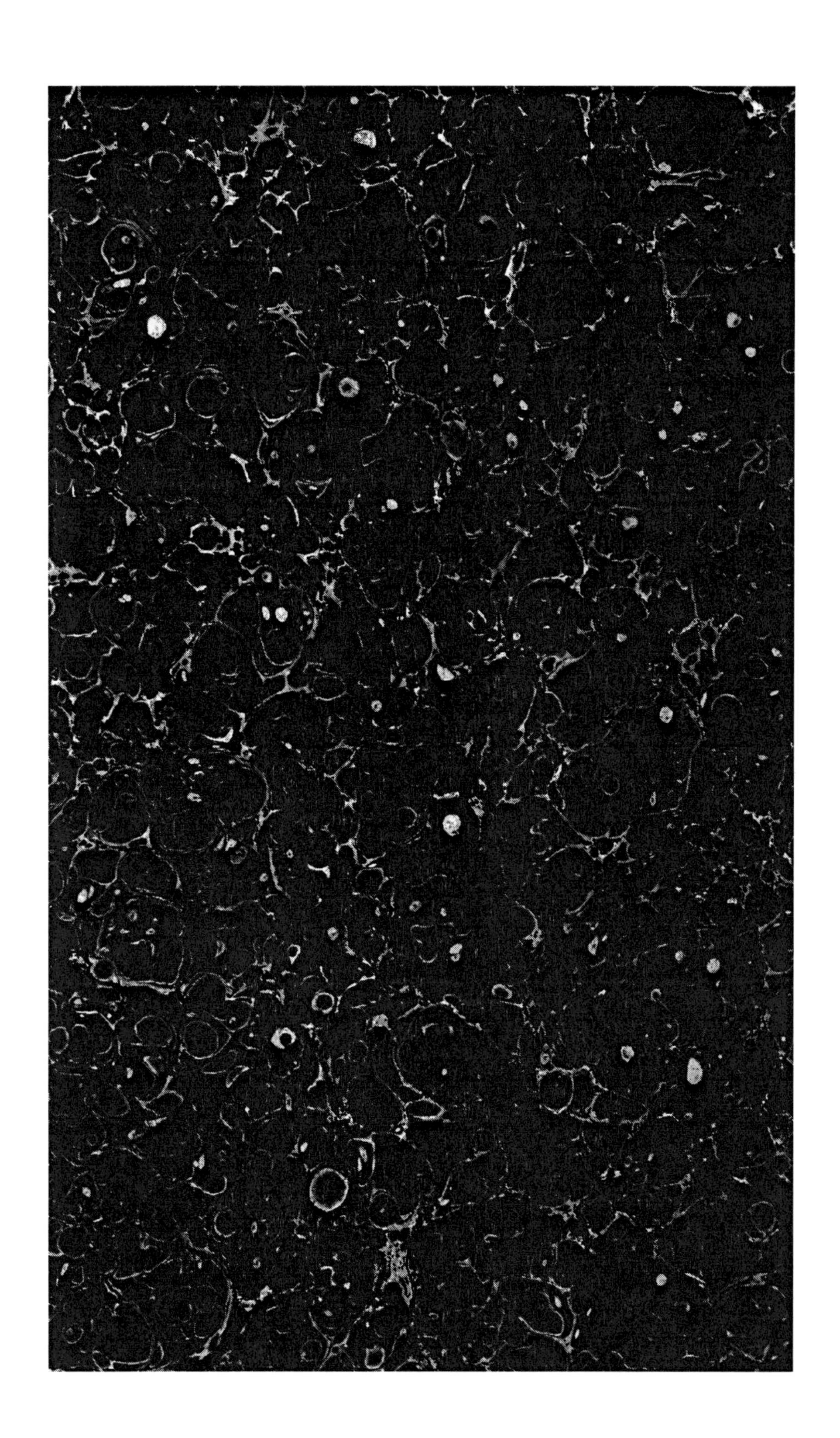

NARRATIVE

OF A SECOND EXPEDITION

TO THE SHORES OF

THE POLAR SEA,

IN THE YEARS

1825, 1826, AND 1827,

BY JOHN FRANKLIN,

CAPT. R. N., F.R.S., &C. AND COMMANDER OF THE EXPEDITION.

INCLUDING AN ACCOUNT OF THE PROGRESS OF A DETACHMENT TO THE EASTWARD,

BY JOHN RICHARDSON, M.D., F.R.S., F.L.S., &c.

SURGEON AND NATURALIST TO THE EXPEDITION.

PUBLISHED BY AUTHORITY OF THE RIGHT HONOURABLE THE SECRETARY OF STATE FOR COLONIAL AFFAIRS.

Philadelphia:

CAREY, LEA, AND CAREY—CHESNUT STREET.

SOLD IN NEW YORK BY G. AND C. CARVILL—IN BOSTON BY MUNROE AND FRANCIS.

1828.

W. PILKINGTON & CO. PRINTERS.

TO

THE RIGHT HONOURABLE

THE EARL BATHURST, K. G.,

LORD PRESIDENT OF HIS MAJESTY'S COUNCIL,

&c. &c. &c.

THE FOLLOWING

NARRATIVE OF A SECOND JOURNEY OF DISCOVERY ALONG THE NORTHERN COAST OF AMERICA,

UNDERTAKEN BY ORDER AND UNDER THE AUSPICES OF HIS LORDSHIP,

IS, BY PERMISSION, INSCRIBED

WITH GREAT RESPECT AND GRATITUDE,

BY

THE AUTHOR.

CONTENTS.

CHAPTER V.

Dr. Richardson's Narrative of the Proceedings of the Eastern Detachment of the Expedition.

CHAPTER I.

CHAPTER II.

CHAPTER III.

CHAPTER IV.

Captain Franklin's Narrative resumed.

CHAPTER VI.

APPENDIX.

An account of the objects of Natural History, collected on our journey, being too voluminous to be inserted in the Appendix, has been reserved for a separate work, which will be published as soon as possible, by Dr. Richardson and Professor Hooker, under the sanction, and by the assistance, of His Majesty's Government.

INTRODUCTORY CHAPTER.

His Majesty's Government having, towards the close of the year 1823, determined upon another attempt to effect a northern passage by sea between the Atlantic and Pacific Oceans, and Captain Parry, the highly distinguished Commander of the two preceding Expeditions, having been again entrusted with its execution, success, as far as ability, enterprise, and experience could ensure it, appeared likely to be the result. Yet, as the object was one for which Great Britain had thought proper to contend for upwards of three centuries, it seemed to me that it might be desirable to pursue it by more ways than one; I therefore ventured to lay before His Majesty's Government a plan for an Expedition overland to the mouth of the Mackenzie River, and thence, by sea, to the north-western extremity of America, with the combined object, also, of surveying the coast between the Mackenzie and Coppermine Rivers.

I was well aware of the sympathy excited in the British public by the sufferings of those engaged in the former overland Expedition to the mouth of the Coppermine River, and of the humane repugnance of His Majesty's Government to expose others to a like fate; but I was enabled to show satisfactorily that, in the proposed course, similar dangers were not to be apprehended, while the objects to be attained were important at once to the naval character, scientific reputation, and commercial interests of Great Britain; and I received directions from the Right Honourable Earl Bathurst to make the necessary preparations for the equipment of the Expedition, to the command of which I had the honour to be nominated.

My much valued friend, Dr. Richardson, offered his services

as Naturalist and Surgeon, and also volunteered to undertake the survey of the coast between the Mackenzie and Coppermine Rivers, while I should be occupied in endeavouring to reach Icy Cape.

Lieutenant Bushnan, who had served under Captains Ross and Parry on their voyages of discovery, was also appointed to accompany me; but, long before the party was to leave England, I had to lament the premature death of that excellent young officer, who was eminently qualified for the situation, by his skill in astronomical observations, surveying, and drawing. Many naval officers, distinguished for their talent and ability, were desirous of filling the vacancy; but my friend and former companion, Lieutenant Back, having returned from the West Indies, the appointment was offered to him, and accepted with his wonted zeal.

Mr. E. N. Kendall, Admiralty Mate, and recently assistant Surveyor with Captain Lyon, was appointed to accompany Dr. Richardson in his voyage to the eastward, and to do the duty of an Assistant-Surveyor to the Expedition at large, whilst it continued united. Lastly, Mr. Thomas Drummond, of Forfar, was appointed Assistant Naturalist, on the recommendation of Professor Hooker, and other eminent scientific men.

A residence in the northern parts of America, where the party must necessarily depend for subsistence on the daily supply of fish, or on the still more precarious success of Indian hunters, involves many duties which require the superintendence of a person of long experience in the management of the fisheries, and in the arrangement of the Canadian voyagers and Indians: we had many opportunities, during the former voyage, of being acquainted with the qualifications of Mr. Peter Warren Dease, Chief Trader of the Hudson's Bay Company, for these services, and I therefore procured the sanction of His Majesty's Government for his being employed on the Expedition.

As soon as I had authority from Earl Bathurst, I entered into a correspondence with the Governor and Directors of the Hudson's Bay Company; and these Gentlemen, taking the most lively interest in the objects of the Expedition, promised their utmost support to it, and forthwith sent injunctions to their officers in the Fur Countries to provide the necessary depôts of provision at the places which I pointed out, and to give every other aid in their power. I also wrote to the different Chief Factors and Chief Traders of the Company, who resided on the route of the Expedition, explaining its objects, and requesting their co-operation.

Pemmican, the principal article of provision used in travelling, being made during the winter and spring, the orders for providing the extra quantity required for the Expedition, though sent out from England by the earliest conveyance, so as to reach the provision posts in the summer of 1824, could not be put into effect sooner than the spring of 1825; hence, it was not proper that the main body of the Expedition should reach the Fur Countries before the latter period. Some stores were forwarded from England, by way of New York, in March 1824, under charge of Mr. Robert M'Vicar, Chief Trader, for the purpose of relieving the Expedition as much as possible from the incumbrance of heavy baggage, and thus enabling it, by marching quickly, to reach its intended winter-quarters at Great Bear Lake, as well as to provide for its more comfortable reception at that place. These stores, with the addition of other articles obtained in Canada, sufficed to load three north canoes, manned by eighteen voyagers; and they were delivered by Mr. M'Vicar, before the winter set in, to Mr. Dease, at the Athabasca Lake. Mr. Dease was instructed to support his party by fishing at Great Slave Lake, during the winter of 1824-25; and, early in the spring of 1825, to proceed to Great Bear Lake, and commence the necessary buildings for the reception of the Expedition. I may here cursorily remark that, in selecting Great Bear Lake as our winter residence, I was influenced by the information I had obtained of its being the place nearest to the mouth of the Mackenzie, known to the traders, where a sufficient supply of fish could be procured for the support of so large a party.

Three light boats, which I shall soon more particularly describe, were also sent out to York Factory, in June 1824, in the annual Hudson's Bay ship, together with a further supply of stores, two carpenters, and a party of men, with a view of their reaching Cumberland House, on the Saskatchawan River, the same season; and, starting from thence as soon as the navigation opened in the following spring, that they might be as far as possible advanced on their way to Bear Lake before they were overtaken by the Officers of the Expedition. The latter, proceeding by way of New York and Canada, would have the advantage of an earlier spring in travelling through the more southern districts; and, further to expedite their progress, I directed two *large* canoes (canôts de maître,) with the necessary equipments and stores, to be deposited at Penetanguishene, the naval depôt of Lake Huron, in the autumn of 1824, to await our arrival in the following spring; having been informed that, in ordinary seasons, we should, by commencing our voyage at that place, arrive in the north-west country ten days earlier than by the usual way of proceeding up the Utawas River from Montreal.

The return of the Hudson's Bay ship towards the close of the year 1824, brought me satisfactory intelligence of the progress of the above-mentioned parties, together with the most pleasing assurances from the Gentlemen of the Company to whom I had written, of their zeal in our cause; and here I must express the deep sense I have of the kindness of the late Honourable William M'Gillivray, of Montreal, whose experience enabled him to give me many valuable suggestions relating to the clothing and subsistence of the party, and to the supplies proper for the Indians.

In connexion with the above sketch of the preparatory steps taken in the course of the year 1824, it may be proper to give, in this place, a short account of the general equipments of the Expedition.

And first, with regard to the vessels intended for the navigation of the Arctic Sea: birch-bark canoes, uniting lightness and facility of repair with speed, are certainly well adapted for navigating the rivers of America, but they are much too slight to bear the concussion of waves in a rough sea, and they are still less fitted, from the tenderness of the bark, for coming in contact with ice. I therefore requested of the Lords Commissioners of the Admiralty that *three* boats might be constructed under my superintendence; and they were immediately ordered and promptly finished under the directions of the Commissioners of the Navy. To fit them for the ascent and descent of the many rapids between York Factory and Mackenzie River; and to render their transport over the numerous portages more easy, it was necessary to have them as small, and of as light a construction as possible; and, in fact, as much like a north canoe as was consistent with the stability and capacity required for their voyage at sea. They were built of mahogany, with timbers of ash, both ends exactly alike, and fitted to be steered either with a sweep-oar or a rudder. The largest, twenty-six feet long, and five feet four inches broad, was adapted for six rowers, a steersman, and an officer; it could be borne on the shoulders of six men, and was found, on trial, to be capable of carrying three tons weight in addition to the crew. The two others were each twenty-four feet long, four feet ten inches broad, and were capable of receiving a crew of five men, a steersman, and an officer, with an additional weight of two and a half tons. The greatest care was paid to their construction by Mr. Cow, boat-builder of Woolwich Yard; and, as I could not often be present, my friend Captain Buchan, R.N., kindly undertook to report their progress; and I am further indebted to him for many valuable suggestions which were acted upon.

When the boats were finished, they were tried at Woolwich, in the presence of many naval and military officers, as to their qualities of sailing, rowing, and paddling, and found to answer fully the expectations that had been formed of them. At the same time we tried another little vessel belonging to the Expedition, named the Walnut-Shell, the invention and construction of which I owe to my friend Lieutenant-Colonel Pasley, of the Royal Engineers. Its length was nine feet, its breadth four feet four inches, and it was framed of well-seasoned ash, fastened with thongs, covered with Mr. Mackintosh's prepared canvas, and shaped like one valve of a walnut-shell, whence its appellation. It weighed only eighty-five pounds, could, when taken to pieces, be made up in five or six parcels, and was capable of being put together in less than twenty minutes. So secure was this little vessel, that several ladies, who had honoured the trial of the boats with their presence, fearlessly embarked in it, and were paddled across the Thames in a fresh breeze. It was intended to provide against a similar detention in crossing rivers to that which proved so fatal to our party on the former journey; and it was also thought, that this little bark would be found useful in procuring water-fowl on the small lakes, to which the boats could not be conveyed.

In the choice of astronomical instruments I was necessarily guided by their portability. Our stock consisted of two small sextants, two artificial horizons, two altitude instruments, a repeating circle for lunar observations, and a small transit telescope for ascertaining the rates of the chronometers. We had a dipping needle mounted on Meyer's plan, a plain needle very delicately fitted for observing the diurnal variation; two of Kater's azimuth compasses, and a pocket compass for each officer. The atmospherical instruments were two electrometers, two of Daniel's hygrometers, Leslie's photometer and hygrometer, besides a good supply of mercurial and spirit thermometers of different sizes. The magnetic instruments were examined in concert with my friend Captain Sabine, previous to my departure from London; and the observations that were obtained for dip and intensity, served as points of comparison for our future results.

The stores consisted of bedding and clothing, including two suits of water-proof dresses for each person, prepared by Mr. Mackintosh, of Glasgow; our guns had the same bore with the fowling-pieces, supplied by the Hudson's Bay Company to the Indian hunters, that is, twenty-eight balls to the pound; their locks were tempered to withstand the cold of the winter; and a broad Indian dagger, which could also be used as a knife, was fitted to them, like a bayonet. Ammunition of the best quality was provided by the Ordnance, the powder being secured in small field or boat magazines. A quantity of wheaten-flour, arrow-root, macaroni, portable-soup, chocolate, essence of coffee, sugar, and tea, calculated to last two years, was also supplied, made up into packages of eighty-five pounds, and covered with three layers of prepared waterproof canvas, of which material coverings for the cargo of each boat were also made.

There was likewise an ample stock of tobacco, a small quantity of wine and spirits, marquees and tents for the men and officers, some books, writing and drawing paper, a considerable quantity of cartridge-paper, to be used in preserving specimens of plants; nets, twine, fishing-lines and hooks, together with many articles to be used at winter-quarters, for the service of the post, and for the supply of our Indian hunters, such as cloth, blankets, shirts, coloured belts, chiefs' dresses, combs, looking-glasses, beads, tapes, gartering, knives, guns and daggers, hatchets, awls, gun-worms, flints, fire-steels, files, whip and hand-saws, ice-chisels and trenching-irons, the latter to break open the beaver lodges.

As the mode of travelling through the Hudson's Bay territories, with all its difficulties and hazards, is now well known to the public, I think it better to give in this Introductory Chapter a slight outline of our route through the United States, Upper Canada, and Southern part of the Fur Countries, and to commence the detailed Narrative of the proceedings of the Expedition with its arrival in Methye River, where the officers joined the boats that had been sent out from England in the preceding year.

On the 16th of February, 1825, I embarked with Lieutenant Back, Dr. Richardson, Mr. Kendall, Mr. Drummond, and four marines, at Liverpool, on board the American packet-ship, Columbia, Captain Lee; and, on quitting the pier, we were honoured by a salute of three animating cheers, from a crowd of the principal inhabitants, who had assembled to witness our departure. The passage across the Atlantic was favourable and pleasant, and our reception at New York kind in the extreme.

We landed at that city on the 15th of March, and our baggage and stores were instantly passed through the Custom-House without inspection. Cards of admission to the Public Scientific Institutions were forwarded to us the same evening, and during our stay every other mark of attention was shown by the civil and naval authorities, as well as by private individuals, indicating the lively interest which they took in our enterprise.

James Buchanan, Esq., the British Consul, in addition to many other attentions, kindly undertook to accommodate a journey he had to make to Upper Canada, so as to accompany us through the State of New York. After a stay of eight days in the city, for the purpose of obtaining the rates of the chronometers, and for making some other observations with Meyer's dipping needle, we embarked under the Consul's guidance, in the steam-boat Olive Branch, and ascended the Hudson River, to Albany, where we experienced similar civilities to those we had received at New York. Every body seemed to desire our success, and a fervent prayer for our preservation and welfare was offered up by the Reverend Dr. Christie, the minister of the church that we attended. The Honourable De Witt Clinton, the Governor of the State, assured me, that had we not been accompanied by a gentleman so conversant in the different routes and modes of travelling as Mr. Buchanan, he would have sent his son with us, or would himself have conducted us to the confines of the State.

From Albany, we travelled through Utica, Rochester, and Geneva, to Leweston, in coaches, with more or less rapidity, according to the condition of the roads; and, crossing the river Niagara, entered Canada, and visited the Falls so justly celebrated as the first in the world for grandeur. We next crossed Lake Ontario in a sailing boat, and came to York the capital of Upper Canada, where we were kindly received by the Lieutenant-Governor Sir Peregrine Maitland, and by Colonel Cockburn and the Commissioners then employed on an inquiry respecting the value of the Crown Lands. From York we passed on to Lake Simcoe, in carts and other conveyances, halting for a night at the hospitable house of Mr. Robinson of Newmarket. We crossed Lake Simcoe in canoes and boats, and landed near the upper part of Kempenfeldt Bay, but not without being obliged to break our way through the ice for a short distance. A journey of nine miles, performed on foot, brought us to the River Nattawassaga, which we descended in a boat; and passing through a part of Lake Huron, arrived at Penetanguishene. At this place, we were hospitably entertained by Lieutenant, now Captain Douglass, during eight days that we waited for the arrival of our Canadian voyagers from Montreal.

We left Penetanguishene on St. George's day (23d April) in the two large canoes, which had been deposited at that place in the preceding autumn, our party, by the accession of the voyagers, now amounted to thirty-three; and after a few days detention by ice, and bad weather, we reached Sault de St. Marie on the 1st of May, being ten days or a fortnight earlier than the oldest resident remembered a canoe from Canada to have arrived. From the Sault de St. Marie, we coasted the northern shore of Lake Superior to Fort William, formerly the great depôt of the N. W. Company, where we arrived on the 10th of May. We now exchanged our two *canôts de maître* for four small north canoes, in one of which, more lightly laden, Dr. Richardson and I embarked, with the view of proceeding as rapidly as possible to arrange supplies of provision at the different posts, while Lieutenant Back was left to bring up the three remaining and more deeply laden canoes.

We proceeded by the route delineated in the maps through Rainy Lake, the Lake of the Woods, Lake Winipeg, and the Saskatchawan River to Cumberland House, where we arrived on the 15th of June, and learned that our boats had left that place on the 2d of the same month. We found also with deep regret, that Thomas Mathews, the principal carpenter who had accompanied the boats from England, had had the misfortune to break his leg the evening before their departure. But, fortunately, an officer of the Hudson's Bay Company then present, had sufficient skill to set it, and Dr. Richardson now pronounced that in two months he would be able to come on in one of the Company's canoes, and join us at Bear Lake, which he was very desirous of doing. I therefore made arrangements to this effect, and also concerning supplies for Mr. Drummond the Assistant Naturalist, who was to be employed, during our stay in the north, in making collections in the vicinity of the Rocky Mountains.

Having remained one night at Cumberland House, we resumed our voyage, and passing through Pine Island Lake, Beaver Lake, crossing the Frog Portage, and ascending the English River, with its dilatations, named Bear Island, Sandfly, Serpent, Primeau, and Isle à la Crosse Lakes, we came to the post situated on, and named from the latter sheet of water, at four P.M. on the 25th June. In the course of this voyage, we met the Gentlemen of the Hudson's Bay Company proceeding from the interior with various brigades of canoes, carrying the returns of trade for the year to York Factory, and I had not only the satisfaction of hearing frequent news of the progress of our boats, but that the deposits of provisions I had requested, and the other arrangements I had made, were all punctually carried into effect. Mr. Spencer, the gentleman in charge at Isle à la Crosse, informed us, that the boats had gone off a few hours previous to our arrival, with the addition of a batteau laden with pemmican, under the charge of Mr. Fraser, a clerk of the Hudsons's Bay Company.

I waited at this establishment one night to obtain astronomical observations, and to bespeak an additional quantity of provisions, &c., which being satisfactorily done, we resumed our voyage on the 27th, and, passing through Deep River, Clear and Buffalo Lakes, overtook the boats in Methye River, at sunrise on the 29th of June.

Having brought this preliminary sketch up to the date at which the ensuing Narrative of the proceedings of the Expedition commences, I turn to the pleasing duty of rendering my best thanks to the many gentlemen who have assisted me in forwarding its progress. To the Right Honourable Earl Bathurst, I am greatly indebted for the readiness with which he attended to every suggestion I had to make regarding the equipment of the Expedition, and to the Right Honourable Wilmot Horton, the Under Colonial Secretary, for his kindness and promptitude in facilitating all my views. Nor can I feel less grateful to Lord Viscount Melville, and to the Lords Commissioners of the Admiralty for their patronage and support, as well as to Sir Byam Martin, the Comptroller, and to the Commissioners of the Navy and Victualling offices, for the arrangements depending on their boards. Mr. Pelly, the Governor of the Hudson's Bay Company, and Mr. Garry, the Deputy-Governor, as well as every Member of its Committee, claim my most sincere thanks for their unremitting endeavours to promote the welfare of the Expedition through its whole progress; and I feel truly obliged to Mr. Simpson, the Governor in the Fur Countries; to Mr. M'Tavish, Mr. Haldane, Mr. M'Donald, Mr. Leith, Mr. Stuart, and Messrs. James and George Keith, Chief Factors, who, acting in the spirit of their instructions, were very assiduous in collecting provisions and stores for the use of my party, and in forwarding all our supplies. There were other gentlemen resident in the more northern parts of the country, to whom I am no less obliged for advice and assistance; but the brevity requisite in this place necessarily compels me to refer to the

Narrative, where their names, and the services they rendered, are mentioned.

I cannot, however, close this introductory Chapter, without expressing the deepest obligation to those kind friends and excellent officers with whom I had the happiness of being associated, who constantly aided me by their most cordial co-operation, and whose best efforts were devotedly applied to every pursuit which could be interesting to science. Nor can I omit to mention the gratitude I owe to each of the seamen, marines, British and Canadian voyagers who composed our party at the winter-quarters, for their steady obedience and truly good conduct, whether in the days of relaxation during the winter, or in the more arduous exertions of our summer occupations.

OFFICIAL INSTRUCTIONS.

Downing-street, 31*st Jan.* 1825.

SIR,

His Majesty's Government having decided that an Expedition should be set forth, for the purpose of exploring the Northern Coast of America, between the Mouth of Mackenzie's River, and the Strait of Behring; and confiding in your zeal and experience for the due execution of this service, I have recommended you as a proper person to be charged with the same. You are, therefore, to proceed with your party (a list of whom is annexed) by the Packet from Liverpool to New York, and from thence make the best of your way to Lake Huron, where the stores necessary for your journey have already been sent. Embarking in Canoes, you are from thence to follow the water communication to the western side of the Great Bear Lake, where you are to establish your winter-quarters; and having so done, your first care should be to endeavour to open a friendly communication with the Esquimaux.

Early in the Spring of 1826, you are to proceed down the Mackenzie River with all the necessary stores and provisions, in order to be prepared to take advantage of the first opening of the ice on the Polar Sea, so as to enable you to prosecute your voyage along the coast to Icy Cape, round which you are to proceed to Kotzebue's Inlet, where you may expect to find His Majesty's Ship Blossom, which the Lords Commissioners of the Admiralty will order to proceed to that rendezvous, in the Summer of 1826. But if, on your arrival at Icy Cape, or the northern point of Behring's Strait, you should be of opinion that you could, with safety, return the same season to the established winter-quarters, you are at liberty to do so, instead of proceeding to join the Blossom. You will, therefore, without loss of time, settle with Captain Beechey, her commander, such a plan as may appear to you, both, best adapted for ensuring your meeting together, and establish a code of signals, or devise such other means as may tend to give you information, if possible, previous to your reaching the longitude of Icy Cape.

On your arrival at the mouth of Mackenzie River, you are to despatch Dr. Richardson with Mr. Kendall and five or six men, in one of the boats, to examine the intermediate coast between the Mackenzie and Coppermine Rivers; but if you should find that the stores and provisions you have been able to accumulate are not sufficient for your own and Dr. Richardson's party, you are, in that case, to direct Dr. Richardson to employ himself and party on shore, in examining the country contiguous to the Mackenzie River, the Rocky Mountains, the shores of the Great Bear Lake, the Copper Mountains, and as far round as he can with safety, collecting specimens of the animals, plants, and minerals, and also laying in a stock of provisions sufficient for both parties, if, by any unforeseen accident, you should find yourself compelled to return without reaching the Blossom.

If, in proceeding westerly towards Icy Cape, you should make but slow progress, and find yourself impeded by ice or land jutting out to the northward farther than is calculated upon, or from accidents to the boats, or any other unforeseen circumstance, so that it remains doubtful whether you will be able to reach the neighbourhood of Kotzebue's Inlet the same season, you are not to consider yourself authorized to risk yourself and party to the chance of being obliged to winter on the coast, but commence your return about the 15th or 20th of August to the established winter-quarters on Bear Lake, unless you should be satisfied that yourself and party could pass the winter with safety among the Esquimaux, and that there was afforded a certainty of your reaching Behring's Strait the following Season, when the Blossom will again proceed to the appointed rendezvous.

In the event of your reaching Kotzebue's Inlet, the first season, Captain Beechey will be instructed to convey you and your party in the Blossom to the Sandwich Islands or Canton, as may seem most advisable to you, from whence you will be able to take a passage to England in one of the Company's Ships or Private Traders; and you will leave such instructions with Dr. Richardson for his guidance, in the event of your being able to accomplish this point, as you may deem fit and proper for his return to England.

In the event of your death, or any accident which may prevent your proceeding, the command of the Expedition must necessarily devolve on Lieutenant Back, who is to follow these Instructions; but he is not to alter any arrangement with regard to Dr. Richardson's proceedings which you may have settled for him to pursue, the principal object of Dr. Richardson's accompanying you, being that of completing, as far as can be done, our knowledge of the Natural History of North America. Lieutenant Back will, therefore, in the event abovementioned, act in concert with Dr. Richardson, but not direct him and his party from any plan of operations which he and you may previously have settled.

You will take care to inform me from time to time, as opportunities may occur, of your proceedings, and the progress made in the Expedition, with the direction of which you are hereby entrusted.

I have the honour to be,
Sir,
Your most obedient Servant,
BATHURST.

To Captain Franklin, R.N.,
&c. &c. &c.

NORTH AMERICA
Scale
P A C I F I C
O C E A N
A T L A N T I C
O C E A N
GULF OF MEXICO
C A R I B B E A N   S E A
G R E E N L A N D
R U S S I A N
Arctic Circle
Tropic of Cancer
S. A M E R I C A
Santa Fe de Bogota

SECOND JOURNEY TO THE SHORES

OF

THE POLAR SEA.

CHAPTER I.

Join the boats in the Methye River—Cross the Long Portage—Arrival to Fort Chipewyan—Departure from thence with the whole party for Mackenzie River—Arrangements at Fort Norman—Descent to the Sea—Return to the Winter Quarters at Great Bear Lake.

THE boats of the Expedition had advanced from Hudson's Bay into the interior, twelve hundred miles, before they were joined by the officers; whilst the latter, from taking a more circuitous route by New York and Canada, as shown in the introductory chapter, travelled two thousand and eight hundred miles, to reach the same point.

June 29.

This junction took place early in the morning of the 29th of June, 1825, in the Methye River, latitude 56° 10′ N., longitude 108° 55′ W., which is almost at the head of the waters that flow from the north into Hudson's Bay.

In no part of the journey was the presence of the officers more requisite to animate and encourage the crews, because the river itself, beside being obstructed by three impassable rapids, is usually so shallow, through its whole course of forty miles, as scarcely to admit of a flat-bottomed bateau floating with half its cargo, much less our boats, which drew, when loaded, from eighteen to twenty inches. This river and its impediments being surmounted, the Methye Portage, ten miles and three quarters long, was at no great distance, which is always held up to the inexperienced voyager as the most laborious part of the journey. But whatever apprehensions the men might have entertained on this subject, seemed to vanish on our landing amongst them; and Dr. Richardson and myself were received by all with cheerful, delighted countenances, and by none more warmly than by our excellent friend and former interpreter Augustus the Esquimaux, and Ooligbuck, whom he had brought from Churchill, as his companion. A breakfast was quickly prepared by Mr. Fraser, a clerk of the Hudson's Bay-Company, under whose charge the boats had been, since their departure from Cumberland House; and I then inspected the boats and stores, which I was rejoiced to find were in good order. We had brought letters from the relatives of several of the party, and another hour was allowed to read them.

At ten A.M. we began to ascend the stream, but very soon found that it was necessary for the whole party to walk in the water, and drag the boats through the mud. Nor could we long advance even by this mode, but were compelled either to carry some of the cargo along the shore, where walking was at all practicable, or else to take half the lading in a boat to a part where the river was deeper, and then return for the remainder. From thus travelling the distance twice over, it was the fifth day before we reached the lake from whence its waters flow.

Thursday, 30. On the evening of the 30th of June, we witnessed one of those violent but momentary gusts of wind which occur not unfrequently in the spring and autumn, and which prove so destructive to the forests in this country. It was preceded by calm and very sultry weather, with loud thunder and vivid lightning. In an instant the tents were overthrown, and even very large trees were bent by its force into a horizontal position; indeed, for a few seconds, the scene around us appeared one of almost entire devastation. When the violence of the squall was past, we had great reason to be pleased at its occurrence, for the strong steady breeze and heavy rain that succeeded, carried away the myriads of musquitoes by which we had been tormented the whole day.

Monday, 4. Having crossed the Methye Lake, we arrived at the portage of that name. Here it was necessary to make an equal division of the cargoes, and to devise means for the conveyance of the boats. The packages amounted to one hundred and sixteen, weighing from seventy to ninety pounds each, exclusive of the three boats and the men's personal luggage; and there were nineteen men of the boats' crews, two Canadians, and two boys, to carry these burdens. At first the packages were equally distributed among this party; but several of the men, who had been reduced by their previous exertions, became lame: among these were the Esquimaux, and we were, therefore, compelled to make other arrangements, and ultimately to employ the crew of my canoe, though the great fatigue they had suffered in our rapid journey from Penetanguishene, made me desirous of sparing them for the present.

The boats were the heaviest and most difficult articles to transport. One of the small boats was carried on the shoulders of eight men, of whom Mr. Fraser undertook to be one, as an example to the rest. Another of the same size was dragged by other eight men; and the largest was conveyed on a truck made for the purpose on the spot, to which service the lame were attached.

Each day's journey, and also the intermediate stages, were determined by the places where water could be procured, and our mode of travelling was as follows:—Rising at three A.M., the men carried a part of their burden to the first stage, and continued to go backwards and forwards till the whole was deposited. They then slept for a few hours, and in the cool of the evening the boats were brought up. By these means every thing was ready at the western end of the portage early on Monday, the 11th of July. The slight injuries which the boats had received, principally from exposure to the sun, were soon repaired; they were put into the water to tighten, and the whole party were allowed to rest.

Monday, 11th.

With reference to the Methye Portage I may remark, that, except the steep hill at its western extremity, the road is good and tolerably level, and it appeared to us that much fatigue and suffering might have been spared by using trucks. Accordingly two were made by our carpenters at Fort Chipewyan, in 1827, for the return of the Expedition, and they answered extremely well. I mention this circumstance, in the hope that some such expedient will be adopted by the Traders for the relief of their voyagers, who have twice in every year to pass over this ridge of hills.

Tuesday, 12th.

Being now in a fair way to reach the Athabasca Lake, Dr. Richardson and I embarked, on the 12th, in the canoe, to proceed to Fort Chipewyan, for the purpose of preparing the gentleman in charge for the reception of the party.

By noon we got over the four Portages on Clear-Water river, and descended, with some trouble, the series of rapids that follow them. Once below these, the passage to the lake is generally considered as free from fatigue; but we did not find it so, for, owing to the shallowness of the water, the men had to get out and drag the canoe in several places. The difference between the depth of water now and in other years at the same period, was attributed to the snow having fallen in the preceding autumn before the frost was sufficiently intense to harden the ground, and, consequently, much of the moisture had penetrated the earth, which, under other circumstances, would have remained in a frozen state, for the supply of the river at the spring thaw.

In the course of the night we were under much alarm for one of our men, who having incautiously lain down to sleep under a wet sail, while the rain was pouring heavily, was seized with a cramp in the stomach, and violent pain in the head. Having been brought into the tent and covered with blankets, he became better before morning, but not sufficiently strong to allow of our setting off at the usual hour. We entered the Elk, or Athabasca River, at three P.M., on the 13th, and were carried swiftly down by its current to the Hudson Bay Company's post named Berens House, where we stayed the night. Here we received a supply of dried meat.

Wednesday, 13th.

We safely arrived in the Athabasca Lake on the 15th, by the channel of the "Riviere des Eaux remuées;" but in the subsequent traverse between Bustard Island and Fort Chipewyan the canoe was in danger of foundering in a sudden gale. Two large waves broke with full force into it, and obliged us to bear away and steer for the nearest shore; but the men having soon rested, and being now sheltered by islands, we pushed on to Fort Chipewyan. Our arrival there caused great surprise to its inmates, when they learned that we had come from England to that advanced post so early in the season, being only two days later than the time at which Dr. Richardson and Mr. Hood had arrived in 1819, though they passed the winter at Cumberland House.

Friday, 15th.

The stores at Fort Chipewyan being well furnished with warm clothing, and other articles, which we required for the use of the men and Indians at winter-quarters, I availed myself of the permission which the chief factor of this department, Mr. James Keith, had given me to complete our stock of cloth, blankets, nets, and twine, to a quantity sufficient for two years' consumption. A supply of twine was indispensable, because, by a letter from Mr. Dease, I had learned that the meshes of the nets made in England, of the size generally required for fishing throughout this country, were too large for the smaller fish that frequent that part of Bear Lake where our house was to be constructed. Mr. Campbell, the clerk in charge, cheerfully gave me the benefit of his experience in making out lists of such things as we were likely to want, and in assorting and packing them.

The boats rejoined on the 18th, and the crews were allowed the following day to recruit themselves. A party of Indians came very opportunely with fresh meat, which is always an agreeable change to the voyager, who has generally to live or dried provision. The Indians, as well as the women and children of the fort, spent the greater part of the day by the side of our boats, admiring their whole equipment, but more especially the gay figures painted on them. Many of these were different from any animals or representations they had seen, and, judging from the bursts of laughter, some curious remarks were made on them.

Monday, 18th.

It being necessary that I should await the arrival of Lieutenant Back's canoes, Dr. Richardson undertook to proceed with the boats towards Slave Lake. Their lading was now increased by the bales already mentioned, as well as by several bags of pemmican, which Mr. Keith had stored up for our use. The crews, however, were reduced by the discharge of three Englishmen, at their own desire, who thought themselves unequal to the fatigue of the service.

Wednesday, 20th.

I had the happiness, on the 23rd, of welcoming my friends, Lieutenant Back and Mr. Kendall, on their arrival with three canoes. Their journey from Fort William had been expeditious, notwithstanding the detention of eighteen days, by bad weather, on the road. A serious misfortune had happened at the very outset of the journey, through the unskilfulness of one of the bowmen, in allowing his canoe to turn round and get before the current, while attempting to ascend the Barrier Rapid, by which it was driven against a stone with such force, as to be overset and broken. The stores were fortunately saved, though completely drenched; but many of the delicate atmospherical instruments were broken. Mr. Kendall was despatched to Fort William for another canoe while the things were drying.

Saturday, 23d.

On a subsequent occasion, in the Winipeg River, the same

man placed his canoe in such a situation, as to endanger its being hurried down a steep fall, and had it not been for the coolness of a man, named Lavallé, who jumped into the water and held the canoe, while the rest of the crew arranged themselves so as to drag it into a place of safety, every life must have been sacrificed. The success, indeed the safety of this kind of river navigation, among currents and rapids, depend on the skill of the bowman; and after these proofs of his incapacity, Lieutenant Back very properly engaged a substitute at the first fort to which he came.

At another time, in the Sturgeon-weir River, the canoe in which Mr. Kendall was embarked, having been accidentally driven before the current, she was only saved from destruction by his own powerful exertion and activity.

These short details will convey an idea of the anxiety and trouble these officers experienced in their journey to Chipewyan.

The party and the stores having now passed the more difficult part of the road, I discharged as many of the Canadians as could be spared, and furnished them with a canoe to take them home. Some went to Montreal; and they were the first persons who had ever gone from that place to Chipewyan, and returned in the same season.

Monday, 25th. The greater part of the 25th was employed in obtaining astronomical observations, the results of which, we were delighted to find, placed Fort Chipewyan within a few seconds of longitude of the position in which it had been laid down on the former Expedition. Our present azimuth compasses showed an increase in variation, since 1820, of 2° 16′ E. The dip was observed 81° 26′ 47″.

Fort Chipewyan was this summer visited, for the first time, by a large flight of swallows, resembling the house-martins of England. They came in a body on the 25th of June, and immediately began to construct their earthy nests under the ledge of the south-front of the house. Some barn or forked-tail swallows also arrived on the 15th of June, and took possession of the store-houses and garrets, as they had in former years done. Some of the young of the last-mentioned birds were sporting on the tops of the houses as early as the 17th of July.

At sunset we embarked in four canoes, one having been procured here. The descent to Slave Lake occupied four days, and was unattended with any circumstance deserving mention, except that two of the canoes were broken in consequence of the guide mistaking the proper channel in a rapid; fortunately, these bark vessels are soon repaired, and we had only to regret the delay the accident occasioned.

We halted at the Salt River to take in salt, as we found, by a note left here, Dr. Richardson had done. The geese were moulting at this time, and unable to fly; they afforded us much sport in their chace, and an excellent supper every night.

A body of Indians were waiting near the entrance of the lake to welcome our arrival; they were so numerous, that we were forced to omit our general custom of giving a small present to each native, and thus incurred the charge of stinginess, which the loud vociferations they raised on our setting sail, were probably meant to convey.

Friday, 29th.

At six, on the evening of the 29th, we reached Fort Resolution, the only establishment now at Slave Lake, and we felt happy in being once more under the roof our hospitable friend, Mr. Robert M'Vicar, to whom I am much indebted for the excellent order in which he had brought up our supplies from Canada in the preceding year. Dr. Richardson, after a halt of two days, had gone forward with the boats.

All the portages on the road to Bear Lake being now passed, the Canadians made a request, that we would allow them to commemorate the event by a dance. It met with a ready compliance; and though they had been paddling for thirty-six out of the thirty-nine preceding hours, they kept up their favourite amusement until daylight, to the music of bagpipes, relieved occasionally by the Jews' harp.

We rejoiced to find at this post our worthy old Copper-Indian friends, Keskarrah and Humpy, the brother of Akaitcho, who had been waiting two months for the express purpose of seeing us. These excellent men showed that their gratification equalled ours, by repeatedly seizing our hands and pressing them against their hearts, and exclaiming, "How much we regret that we cannot tell what we feel for you here!" Akaitcho had left the fort about two months on a hunting excursion, hoping to return, with plenty of provision for our use, by the middle of August, which was as early as he thought we should arrive. Keskarrah confirmed the melancholy report we had heard in the more southern districts, that most of the hunters who had been in our service at Fort Enterprise, had been treacherously murdered, with many others of the tribe, by the Dog-ribs, with which nation we also learned the Copper-Indians had been at war, since the year of our departure from them, till the last spring. The peace had been effected through the mediation of Messrs. Dease and M'Vicar, and we were gratified to find that Akaitcho and his tribe had been principally induced to make this reconciliation, by a desire that no impediment might be placed in the way of our present expedition. "We have too much esteem," said Akaitcho, "for our father, and for the service in which he is about to be again engaged, to impede its success by our wars, and, therefore, they shall cease;" and on being asked by Mr. Dease whether he and some of his young men would go to hunt for the party at our winter quarters, he replied, "Our hearts will be with them, but we will not go to those parts where the bones of our murdered brethren lie, for fear our bad passions should be aroused at the sight of their graves, and that we should be tempted to renew the war by the recollection of the manner of their death. Let the Dog-ribs who live in the neighbourhood of Bear Lake furnish them with meat, though they are our enemies." Such sentiments would do honour to any state of civilization, and show that the most refined feelings may animate the most untutored people. Happily we were now so circumstanced as to be able to reward the friendship of these good men by allotting from our stores a liberal present to the principal persons. On the delivery of the articles to Keskarrah and Humpy, I desired them to communicate to Akaitcho, and the whole tribe, the necessity of their strictly adhering to the terms of peace, and assured them that I should not fail to urge the same obligations on the Dog-ribs. A silver royal medal, such as is given to the Indian chiefs in Upper Canada, was likewise left with Mr. M'Vicar, to be presented to Akaitcho, as a further mark of our regard for his former services and present good wishes.

Sunday, 31st. The party was detained at Fort Resolution until this morning by a strong south-west gale; and even when we embarked, the wind and waves were still high, but time was too precious to allow of our waiting when there was a prospect of making any advance. As our future course inclined to the westward, we now quitted the track of the former journey to Fort Enterprise, along which we had been travelling from Lake Winnipeg. We first steered for the Buffalo River, and then along the south shore of Slave Lake, obtained the latitude 61° 1′ N. at noon, and afterwards the longitude 114° 18½′ W. at the Isle of the Dead. The islands and shores of this part of the lake are composed of horizontal beds of limestone, containing pitch and shells.

A small party of Chipewyan Indians, with their principal chief, joined us at the encampment, from whom we learned that they had supplied Dr. Richardson with dried meat the preceding noon, at Hay River. The Chief was very importunate for rum, but I steadily adhered to the determination I had formed this time, on my entering the Fur Country, of not giving spirits to any Indian. A share of our supper and tea, and some tobacco, were offered to him, and accepted, though with a bad grace. The Fur Company ceased the following season to bring any rum to this quarter, and I learned that this man was one of the few natives who were highly displeased at this judicious change.

Monday, 1st.

We coasted this day along the low shore of the lake, steering from point to point to avoid the sinuosities of several deep bays, and passed the mouth of the Sandy and Hay Rivers, whose positions we settled by astronomical observations.

Tuesday, 2nd.

On the 2nd we came to the narrow part between the Big Island of Mackenzie, and the main shore, and perceived that a gentle current was setting towards the Mackenzie river. The water in this strait is very shallow, and also in many places near the south shore, though we know, from trial, on the former Expedition, that the depth of the east end of the lake, at a distance from the land, exceeds sixty and seventy fathoms. The beach, both of the north and south shores of the strait, is strewed with drift timber. In clear weather the north shore is visible from the point of the south shore nearest Big Island.

Below this *detroit* the shores recede so as to form a small shallow lake, about twenty-four miles long, by from four to twelve miles broad, near the north-west end of which we encamped, in latitude 61° 15′ N., longitude 117° 6′ W. This spot may be considered as the commencement of Mackenzie River. The ground is very swampy, and nourishes willows only; but inland, at a short distance from the beach, grow plenty of the spruce-fir, poplar, aspen, and birch trees; and among the underwood, numerous shrubs and berry-bearing plants.

Wednesday, 3rd.

On the 3rd we travelled to another contraction of the river about one mile broad, through which the current sets between high banks with such force as to form strong eddies. There are likewise in this part many sandy islands, and through the channels between them the current rushed with no less rapidity than in that we descended. For distinction's sake, these islands have been named the "Isles of the Rapid:" below them occurs another expansion, which is called by the voyagers "The Little Lake;" and Sandy Point at its northwest end, is considered by them as the commencement of the Mackenzie River.

When abreast of this point, a favouring breeze enabled us to use the sail as well as the paddles, and with the assistance of the current great progress was made. We had occasional glimpses of the Horn and Rein-Deer Mountains as we passed along; but, until we were some way below the rapids, our view was very limited, owing to the woods being on fire in almost every direction. This I should have mentioned to have been the case in many parts between Isle à la Crosse and the Mackenzie. The cause of these extensive conflagrations I could not learn; some attributed them to voluntary acts of the Indians, and others to their negligence in leaving their fires burning.

We put up at sunset on a beach of gravel under a well-wooded bank of moderate height, and the party regaled themselves with raspberries and other indigenous fruits.

Thursday, 4th.

At half past two A.M., on the 4th, the canoes were again on the water, and being driven by sail and current, made good way. We stopped at the Trout River, which flows in from the southward, and ascertained its longitude 119° 47′ W. The breadth of the Mackenzie is here about two miles, and its banks are composed of a muddy clay: the stones on the beach mostly limestone, with some boulders of primitive rocks. The trees are of the kinds we had seen north of the Athabasca Lake: they are here of a smaller size. Five miles below this part, the Mackenzie is divided into several channels by islands, and the current runs with increased swiftness, and strong eddies.

The latitude 61° 26′ 30″ N. was obtained at noon; it was the same as on the preceding day; so that our course, in the interval, had been due west.

The banks now were higher, and for the next forty miles the breadth of the stream did not exceed one mile, nor was less than half a mile; its course inclined more to the north. We passed the site of the first establishment that the North-West Company had made in these parts, which was erected by Mr. Livingstone, one of the partners, who, with the whole of the crew of his canoe, except one individual, were massacred by the Esquimaux on the first attempt to open a trade with them.

At three P.M. a picturesque view opened upon us of a distant range of mountains running east and west, and nearly at right angles to the course of the river. The current being considerably increased by the contribution of some streams near this place, we descended very swiftly. Six miles below Pine Island, there is a strong but not a dangerous rapid; and about fifteen miles farther is Fort Simpson, the principal depôt of the Hudson Bay Company for this department, at which we arrived by eight P.M., and thus escaped a very wet, comfortless night. Dr. Richardson had departed for Fort Norman the preceding day.

This establishment, three hundred and thirty-eight miles from Fort Resolution on Slave Lake, is situated at the confluence of the River of the Mountains and the Mackenzie. The former is the channel of communication with a fur post not far distant from the Rocky Mountain Range, from whence the residents here procure much of their provision, including a tolerable supply of potatoes, which have been recently introduced from the southern parts. Mr. Smith, the chief factor of the district, was fortunately at Fort Simpson, so that I had the opportunity of arranging with him as to supplies of provision or stores that my party might require during its residence at Bear Lake. He cheerfully acceded to every suggestion that was made, and likewise furnished me with a letter of instruction to the same effect, addressed to the gentleman in charge of the lower posts.

I learned from Mr. Smith that, as yet, a few only of the Indians who live nearest the mouth of the river, and none of the Esquimaux, had been apprized of our intended visit, the traders at the lower posts having considered that it would be better to defer this communication until we should arrive in the river, for fear of disappointing these people, which might have been attended with unpleasant results.

There were two Canadians here belonging to the Expedition, whom Mr. Dease had sent to serve as guides to Bear Lake. By letters which they brought, I was informed that Indian hunters were engaged, and the necessary buildings in course of preparation for our reception. As Fort Simpson had been short of ammunition during the summer from some accidental cause, I was glad to find that Mr. Dease had been enabled to lend from our stores a barrel of powder, and a bag of balls and I now increased the loan, so as to meet the probable demands of the Indians, until the Company's supplies should arrive, when they would return to Fort Norman the whole of what we had lent. Cloudy weather limited our astronomical observations at this place to the dip of the needle, which was observed 81° 54′.

Friday, 5th. We quitted the fort on the 5th, soon after noon, whence the river preserving nearly a straight course for fifteen miles, gradually extends itself to nearly two miles in breadth: in its channel there are three islands. At two P.M. we obtained the first glimpse of the Rocky Mountains, and kept them in view until we encamped, which was early, as the canoes required gumming. The outline of the mountains was very peaked, and at their eastermost part was a cone-shaped hill, higher than the rest, whose summit was veiled by clouds. The general appearance of the range somewhat reminded me of the east end of Jamaica.

Saturday, 6th. The morning of the 6th was beautifully fine: we embarked at 2_h 30′ A.M., and by seven came within six or seven miles of the mountain range, where the river suddenly changes its course from W.b.N. to north, in longitude 123° 31′ W.

A distinct stratification was perceptible on the face of the nearest mountain: on one side of a nearly perpendicular ravine the strata dip to the southward at an angle of 25°; whilst on the other they are nearly horizontal. There was a large accumulation of debris at its base: every part of the hill was destitute of vegetation. Its altitude was guessed at one thousand two hundred feet.

At noon, in latitude 62° 49′ N., we saw a chain of mountains, on the eastern side of the river, similar in their outline and general character to those hitherto seen only on the opposite bank. Between these ranges the river flowed in a channel two miles broad; but as we advanced we receded from those on the western side, their direction being W.N.W. In the brilliancy of the sunshine, the surfaces of some of the eastern hills, which were entirely bare, appeared white as marble, and for some time we fancied them to be covered with snow. By four P.M. we reached the Rocky Island mentioned by Mackenzie, where, from the river being contracted, the current flowed with great rapidity, and soon brought us opposite to the remarkable hill close by the river side, which that persevering traveller ascended in July, 1789. His account renders a description of it unnecessary. It is composed of limestone, and is about four hundred feet high.

We continued a N.b.W. course for eight miles, and encamped at sunset, having travelled this day one hundred and twenty miles. A small supply of fresh deer's meat was obtained from some Dog-Rib Indians. Their canoes were made of the bark of the pine-tree, sewn at the ends and top with the fibrous parts of the root of that tree, leaving only a space sufficient for the legs of the sitter.

Sunday, 7th.

We pursued our course at dawn of day, and at the end of a few miles came to a more winding part of the river, where the stream is interrupted by numerous sand banks and shoals which we had some trouble to get round. Mr. Kendall, in his Journal, remarks of this part, "That bubbles of air continually rose to the surface with a hissing noise resembling the effervescence produced by pouring water on quick lime."

We arrived at Fort Norman at ten, A.M., distant two hundred and thirty-six miles from Fort Simpson, and five hundred and seventy-four from Fort Resolution.

Being now only four days' journey from Bear Lake, and there remaining yet five or six weeks of open season, I resolved on following up a plan of a voyage to the sea, which I had cherished ever since leaving England, without imparting it to my companions, until our departure from Fort Chipewyan, because I was apprehensive that some unforeseen accident might occur in the course of the very intricate and dangerous river navigation between Fort William and the Athabasca Lake, which might delay our arrival here to too late a period of the year. It was arranged, *first*, that I should go down to the sea, accompanied by Mr. Kendall, and collect whatever information could be obtained, either from actual observation, or from the intelligence of the Loucheux Indians, or the Esquimaux, respecting the general state of the ice in the summer and autumn; the direction of the coast, east and west of the Mackenzie; and whether we might calculate upon any supply of provision. *Secondly*, Dr. Richardson, on his own suggestion, was to proceed in a boat along the northern shore of Bear Lake, to the part where it approached nearest to the Coppermine River, and there fix upon a spot to which he might bring the party the following year, on its return from the mouth of that river. And, *thirdly*, that these undertakings might not interfere with the important operations necessary for the comfortable residence and subsistence of the Expedition during the following winter, Lieutenant Back was to superintend them during my absence, with the assistance of Mr. Dease, chief trader of the Hudson Bay Company, whose suggestions, relative to the proper distribution of the Indian hunters, and the station of the fishermen, he was to follow. Accordingly, Dr. Richardson, on his quitting this place two days previous to our arrival, had left the largest of the boats, the Lion, for my use and a well-selected crew of six Englishmen, and Augustus the Esquimaux.

Lieutenant Back was directed to take the canoes forward to Bear Lake, laden with such supplies as would be required for the winter, and was further instructed to furnish Dr. Richardson with one of the boats, and a good crew. The services of the Canadians who had brought the canoes from Penetanguishene, being no longer required, I desired Lieutenant Back to discharge them, and also all the voyagers of Mr. Dease's party who could be spared. They were sent in canoes to Slave Lake, where I had arranged with Mr. M'Vicar for their being supplied with the means of gaining subsistence by fishing, during the winter; and the following spring, they were to be forwarded to Canada, at the expense of Government, according to the terms of their agreement.

Fort Norman being situated in our way to the sea, the pemmican and other stores, intended for the voyage along the coast next season, were deposited here, by permission of Mr. Smith, under the care of Mr. Brisbois, the clerk in charge. Our observations place this establishment in latitude 64° 40′ 30″ N., and longitude 124° 53′ 22″ W.

Monday, 8th. The above matters being satisfactorily settled, and a few articles packed up as presents to the Indians and Esquimaux, Mr. Kendall and I embarked on the 8th, at noon, taking, in addition to our crew, a voyager, who was reported to be able to guide us through the proper channels to Fort Good Hope, of which, however, we found him altogether ignorant. We were accompanied by Lieutenant Back, with the three canoes, each manned by five men. The crews of the canoes imagining they could easily pass our English boat, were much surprised, on putting it to the proof, to find the boat take and maintain the lead, both under sail and with oars.

A few miles above the Bear Lake River, and near its mouth, the banks of the Mackenzie contain much wood coal, which was on fire at the time we passed, as it had been observed to be by Mackenzie in his voyage to the sea. Its smell was very disagreeable. On a subsequent trial of this coal at our winter quarters, we found that it emitted little heat, and was unfit for the blacksmith's use. The banks likewise contain layers of a kind of unctuous mud, similar, perhaps, to that found on the borders of the Orinoco, which the Indians, in this neighbourhood, use occasionally as food during seasons of famine, and even, at other times, chew as an amusement. It has a milky taste, and the flavour is not disagreeable. We used it for whitening the walls of our dwellings; for which purpose it is well adapted.

The entrance of the Bear Lake River is distinguished by a very remarkable mountain, whose summit displays a variety of insulated peaks, crowded in the most irregular manner. It is composed of limestone; and from the lower cliffs, which front the river, a dark, bituminous liquid oozes and discolours the rock. There are likewise two streams of sulphureous water that flow from its base into the Mackenzie. At this place we parted from our friend, Lieutenant Back, who entered the clear and beautiful stream that flows from Bear Lake, of whose pure waters we had also the benefit, till they were overpowered by the muddy current of the Mackenzie. The day was fine, the wind fair, the current swift, and every circumstance concurred to put the party in high glee. There was little in the scenery to attract our attention, now that we had become familiar with the general appearance of the Mackenzie, and we passed island after island, of the same alluvial mud, without further regard than the delineation of them in the survey book. At length, however, a most picturesque view of the Rocky Mountain range opened before us, and excited general admiration, and we had also some portions of the mountain range on the eastern side of the river, in view for the remainder of the day's journey. The outline of these mountains is very irregular, the highest parts being peaked hills. The general direction of the ranges is between N.W. and N.W.b.W.

Being unwilling to lose the advantage of the wind, we only put ashore to sup, and after two hours' delay, resumed our voyage under easy sail. When the sun rose, the oars were used; and then, as the current set at the rate of two miles and a half per hour, the boat travelled swiftly down the stream. The eastern bank of the river, along which we were passing, is about one hundred and twenty feet high, almost perpendicular, and is composed of thin strata of bituminous shale. Amongst the fragments of shale which strewed the beach, we found many pieces of brown wood-coal. A reach, eighteen

Tuesday, 9th. miles in length, followed. It is bounded on both sides by high cliffs of sand-stone. We landed to breakfast, and to obtain the longitude, 128° 23′ W.

From the reach here described, are seen two hills, named by me the East and West Mountains of the rapid, which seem to present a barrier to the further progress of the stream; but the river, bending suddenly between them to the north, dilates into a kind of basin, and, by so doing, opens by far the most interesting view of the Rocky Mountains which the Mackenzie affords. The river, too, makes its nearest approach to those mountains at this spot, and, probably, the easiest communication with them would be by ascending a small stream that flows in here on the western side. Here too are found the first rapids mentioned by Mackenzie, which continue in succession for two miles, when the water is low. The centre of the basin is occupied by low sandy islands; and the channel on the western side is the deepest. The beauty of this scene furnished employment for the able pencil of Lieutenant Back, on a subsequent occasion. As the Mackenzie, in its further descent, continues to hold a northerly course, and the range of mountains runs N.W.b.N., we did not obtain any other view of them till we approached the sea.

At one P.M. we saw a party of Indians encamped on the beach of a small stream, whom we invited to come off to us. They hesitated at first, being doubtful who we were, from our boat being different in shape from any they had seen, and carrying two sails; but after some time they launched their canoes, and brought us a good supply of fresh deer's meat. The sight of our boats seemed to delight them as much as the ammunition and tobacco which they received. These were Hare Indians, the tribe that follows next to the Dog-Ribs, in the line of country below Bear Lake; and, like them, they speak a dialect of the Chipewyan language. We admired the shape and appearance of their canoes, which were larger than those used by the Chipewyans, and had the fore part covered with bark, to fit them for the navigation of this broad river, where the waves are often high.

The river varied from two to four miles in breadth, and its course was interrupted by several small islands and sand-banks. At six P.M. we came to an open space, bounded by lofty walls of sandstone. In this expansion are found the second rapids of Mackenzie: at the first appearance they seem dangerous, but are not so. The river becomes again contracted, and rushes with great force for the space of seven miles through a kind of defile, varying in breadth from four hundred to eight hundred yards, which has been appropriately named "The Ramparts," by the traders. The walls of this defile are from eighty to one hundred and fifty feet high, and are composed of limestone, containing numerous shells: for a part of the way the stone is very white, and in the rest it is blue. Several streams of water were running over the summits of the cliffs, which had worn the stone, in some places, into a turreted shape; while the heaps, overthrown by its action at their base, resemble mounds for defence. To these appearances were occasionally added cavernous openings, and other hollow parts, not unlike the arched windows or gateways of a castellated building. I could not help fancying what delight a visit to this spot would afford to any person of a romantic turn, especially at the time we first saw it, when the broad shadows of a declining sun gave effect to the picture. This is a place of resort for the Hare Indians to fish, and we were visited by a large party of men and women of that tribe, who brought fish, berries, and meat. They were all neatly clothed in new leathern dresses, highly ornamented with beads and porcupine quills. The paintings of animals on the sides of our boats were very attractive to them; they scanned every figure over and over, bursting into laughter whenever they recognised any of the animals. We encamped near a small river below the ramparts, one hundred and ninety-three miles from Fort Norman. Two young Indians followed us in their canoes, bringing some musk-rat skins, and fish for sale. We purchased the fish, but declined taking the furs. They were so pleased with their reception, that they passed the night by our fire.

Wednesday, 10th.

At day-light we again embarked, and descended the river pleasantly and swiftly under sail, having the benefit of a strong current, especially where it was narrowed by islands or sand-banks. The sides of the river are generally high cliffs of limestone or sandstone, and its breadth from two to three miles. The intervals between these cliffs are mostly occupied by hills of sand, from eighty to one hundred feet high, whose intermediate valleys are well wooded; and whenever these occur, the channel of the river is much interrupted by banks, on which, as well as on the beach, there are vast collections of drift timber, piled, in some places, twenty feet high, by the spring floods.

At eleven P.M. we arrived at Fort Good Hope, the lowest

of the Company's establishments; it is distant from Fort Norman three hundred and twelve miles, and is in latitude 67° 28′ 21″ N., and longitude 130° 51′ 38″ W.: the variation of the compass being 47° 28′ 41″ E. Our arrival at this period of the year, at least two months earlier than that of the Company's boats from York Factory, caused great astonishment to the few inmates of this dreary dwelling, and particularly to its master, Mr. Charles Dease, who scarcely recovered from his surprise until we had been seated some time in his room. But this over, he quickly put every one in motion to prepare a meal for us, of which we stood in much need, as it was then verging on midnight, and we had breakfasted at eight in the morning. This post had been but recently established for the convenience of the tribe of Indians whom Mackenzie calls the Quarrellers, but whom the traders throughout the fur country name Loucheux. As this name is now in general use, I shall adopt it, though it is but justice to the people to say, that they have bright sparkling eyes, without the least tendency to that obliquity which might be inferred from the term. The fact is, that Loucheux, or Squinter, was intended to convey the sense of the Indian name of the tribe—Deguthée Dennee, which means "the people who avoid the arrows of their enemies, by keeping a look out on both sides." None of the tribe was at this time at the fort; but from Mr. Dease we learned the interesting fact, that the Loucheux and Esquimaux, who are generally at war, had met amicably the preceding spring, and that they were now at peace. We procured from the store an assortment of beads, and such things as were most in request with the Loucheux, and made up a small package of clothing to be presented to each chief of that tribe, whose favour it was thought advisable by this means to propitiate, as they were the next neighbours to the Esquimaux.

Thursday, 11th. After the latitude had been observed, we embarked, and were accompanied by Mr. Dease as far as Trading River, where he expected there might still be a party of Indians, which did not prove the case. This river being the usual limit of the trader's travels towards the sea, the voyager who had come with us from Fort Norman declined going any farther, and by permission of Mr. Dease he was exchanged for a young half-breed named Baptiste, the interpreter of the fort, who went under the promise of being left with the chief of the Loucheux, to whom he was to introduce the party.

The reach below Trading River is remarkable, from the banks on the eastern side consisting of hills of a light yellow marl-slate, nearly uniform in shape, and strongly resembling piles of cannon shot. The name of Cannon-Shot Reach was, therefore, bestowed on it. The channel of the river is very intricate, winding amongst numerous sand-banks, and some low alluvial islands, on which willows only grow. Its breadth is about two miles, and the depth of water, in the autumn, from six to twelve feet. In passing through Cannon-Shot Reach, we were hailed by an Indian from the shore, and landed immediately, to inform him of the purport of our visit. As soon as Baptiste had explained these matters to him, the man, deeming it of importance that we should be properly introduced to his relatives, offered to accompany us to the next party, providing we would undertake to carry his baggage. This we consented to do, little expecting, from the appearance of poverty in himself and his family, and still less from that of his tent, a mere covering of bark and pine branches, supported on three poles, that load upon load of unsavoury fish would be tossed into the boat. However, we were unwilling to retract our promise, and suffered our vessel to be completely lumbered. We then pushed off, leaving the family to follow in the canoe, but in a short time our ears were assailed by the loud cries of the man demanding that we should stop. On his coming up, we found he was apprehensive of the canoe sinking, it being very leaky and overloaded, and of his losing his wife and infant child. The water being thrown out, the man proposed going forward and keeping by our side. There was nothing now to fear, yet the lamentations of the woman became louder and louder, and at last the poor creature threw off her only covering, raised the most piteous cries, and appeared a perfect object of despair. We learned from Baptiste that she was mourning the loss of two near relatives who had recently died near the spot we were passing. In this manner do these simple people show their sorrow for the death of their connexions. As we drew near the tents of the party on shore, the husband proclaimed with a stentorian voice who we were; this produced a long reply, of which Baptiste could only collect enough to inform us that many persons were lying sick in the lodges, and that two had died the preceding day. Not choosing to expose ourselves to the hazard of contagion, we put the baggage of our friend on shore at some distance below the lodges. All those who were able to manage a canoe, came off to receive presents, and to see

Augustus, the principal object of attraction. Each person crowded to the side on which he sat to shake him by the hand; and two of the party, who had been occasionally with the Esquimaux, contrived to make him understand that, being accompanied by him, we need apprehend no violence from them, though they were a treacherous people. At the end of five miles farther we put on shore to sup, and afterwards slept in the boat; but Augustus spread his blankets on the beach before the fire, and allowed four of the Loucheux, who had followed us from the tents, to share them with him.

Friday, 12th. At daylight we loosened from the beach, and continued with the descent of the river; winding, in our course, as numerous sandbanks rendered necessary. In a few hours we descried another collection of Indian lodges. One of the party happened to be examining his nets nearer to us than the tents; on espying the boat, he immediately desisted, and paddled towards his friends with the utmost speed, bawling the whole way for them to arm. The women and children were seen hurrying up the bank to hide themselves; and by the time we had got abreast of the lodges, the whole party were in a state of defence. They stood on the beach gazing at us evidently with much distrust; and for some time no one would accept our invitations to approach. At length an adventurous youth, distinguishable among the rest by the gaiety of his dress, and the quantity of beads that were suspended around his neck, launched his canoe and paddled gently towards the boat, till he discovered Augustus, whom he knew by his countenance to be an Esquimaux; then rising from his seat, he threw up his hands for joy, and desired every one of the party to embark at once. The summons was instantly obeyed, and a friendly intercourse followed; each person that had a gun discharging its contents, and taking the iron heads and barbs from the arrows, to show their entire confidence. On landing to breakfast, we found that the dialect of this party was different from that of the men we had seen yesterday, and that Baptiste did not understand their language; consequently our communications were carried on by signs, except when they attempted to speak Esquimaux, which Augustus, with difficulty, made out. He was still the centre of attraction, notwithstanding Mr. Kendall and myself were dressed in uniform, and were distributing presents to them. They caressed Augustus, danced and played around him, to testify their joy at his appearance among them, and we could not help admiring the demeanour of our excellent little companion under such unusual and extravagant marks of attention. He received every burst of applause, every shake of the hand, with modesty and affability, but would not allow them to interrupt him in the preparation of our breakfast, a task which he always delighted to perform. As soon as we had finished our meal, he made his friends sit down, and distributed to each person a portion of his own, but without any affectation of superiority. When we were on the point of embarking, the oldest Indian of the party intimated his desire that we should stop until some one whom he had sent for should come. This proved to be his son, in a very sickly state. Though the day was warm, the lad was shivering with cold, and it was evident he was suffering from fever, which the father had no doubt we could cure. The only remedy we could apply was some warm tea, with a little brandy in it, which we afterwards learned had the desired effect of restoring the invalid. Again we were preparing to set off, when the same old man begged us to stop until the women should come; these were no less pleased with Augustus, and with the presents they received, than the men had been.

This good-natured tribe is distinguished by the traders as the Lower Loucheux, but the literal meaning of their Indian name is the Sharp Eyes. They are decidedly a well-looking people: in manner, and general appearance, they resemble the Esquimaux near the mouth of the Mackenzie, though not in their eyes, which are prominent and full. Their canoes, too, are shaped like those of the Esquimaux, and made of birch bark, which, by some process, is striped from the gunwale perpendicularly downwards, for the purpose of ornament. Their summer dress, like that of the Upper Loucheux and Esquimaux, is a jacket of leather, prolonged to a point before and behind: the leggings, of the same material, are sewn to the shoes, and tied by a string round the waist. The outer edges of their dress are cut into fringes, coloured with red and yellow earth, and generally decorated with beads. Beads are so much coveted by them, that, for some years, they were the principal article of trade exchanged for their furs; and even now the successful hunter, or the favourite son, may be known by the quantity of strings of different coloured beads which he has about his neck. These Indians are the only natives of America, except the Esquimaux, whom I have seen with the septum of the nose perforated, through which, like the Esquimaux, they thrust pieces of bone, or small strings of shells, which they purchase from that people. Few of them have guns, but each man is armed with a bow and arrows. The bows are constructed of three pieces of wood, the middle one straight, and those at each end crooked, and bound with sinews, of which the string is also made. The dress of the women only differs from that of the men by the hood being made sufficiently wide to admit of their carrying a child on their back.

At ten A.M. we resumed our journey, followed by the young man who had first spoken to us, and his brother, in their canoes, and in the course of two hours came abreast of a remarkable round-backed hill, on which we were informed Mr. Livingstone and his party had encamped in 1795, the night before they were massacred. This hill marks the commencement of another contraction of the river, which is here pent in between very steep cliffs of blue limestone, which I have denominated the Narrows. The Red River contributes its waters to the Mackenzie at the lower part of the Narrows, in latitude 67° 27′ N., longitude 133° 31′ W.; and, though of inconsiderable size, is remarkable as being the boundary between the lands claimed by the Loucheux Indians and those of the Esquimaux, and likewise as the spot where the amicable meeting between these tribes had been held in the preceding spring. We did not find the chief of the Loucheux here, as had been expected, and therefore passed on. The banks of the river, now entirely composed of sand and sandstone, became gradually lower, and more bare of trees. At the end of eight miles we arrived at a very spacious opening, in which were numerous well-wooded islands, and various channels. The rocky mountains on the west once more appeared in view, extending from S.W. to N.W. and preserving a N.W.½W. direction; and of this range a very lofty peak, and a table mountain, which I have named after the late Mr. Gifford, form the most conspicuous features. We steered into the eastern channel, as being that through which the current seemed to run swiftest; and as soon as we came to a high bank we landed, for the purpose of taking a survey of the surrounding scene. But even from its summit our view was very limited, and all we could discover was, that we were certainly in that expansion of the river that Mackenzie delineates in his chart, and, therefore, in the fair way to the sea, whatever channel we took. This might have been inferred, from the sudden departure of our two Indian companions, who dropped behind and turned their canoes round, without further ceremony, as soon as they saw our intention of entering the eastern channel. Baptiste, who was asleep at the time, expressed surprise at their having gone back, but consoled himself with the idea of meeting the Indian chief the next morning, at a place he called the Forks. We were amused at conjecturing how great his surprise would be should he next be disturbed by the hallowing of a party of Esquimaux, whom he greatly dreaded. At the end of twenty-three miles descent in the middle channel, having passed one that branched off to the eastward, we put up at an early hour, and caused the guns to be cleaned, and two sentinels appointed to watch, least the Esquimaux should come upon us unawares. The banks of the river, as well as the islands, are entirely alluvial, and support willows at the lower parts, and the spruce-fir trees at the summits. The beach on which we were encamped was much intersected with the recent tracts of the moose and rein-deer.

Saturday, 13th.

We embarked at three A.M. on the 13th; and as we were in momentary expectation of meeting the Esquimaux with whom I wished to have an interview, the masts were struck, lest they should discover the boat at a distance, and run off. We soon passed two of their huts, which did not seem to have been recently inhabited. The longitude 134° 20′ 30″ W., and variation 51° 4′ 20″ E., were observed at the time we halted to breakfast, and the latitude 68° 15′ 50″ N., at noon. The Rein-deer mountains on the eastern side, came in view before noon. The range on the west was also occasionally visible: we were descending between the M'Gillivray and Simpson islands, in a channel that did not exceed half a mile in breadth. A fine breeze sprung up after noon, of which we took advantage by setting the sails, not having seen any recent traces of the Esquimaux. At the extremity of Simpson island there is a broad channel, which pours its waters into the one in which we were, at a place where the stream is contracted by a small island, and a strong rapid is the consequence of this junction. Here we found many huts, and other indications of its being a place of resort for fishing; here, too, it is supposed Mr. Livingstone and his crew fell a sacrifice to the first party of Esquimaux whom they met. Several other openings branched off to the eastward; but we continued to follow the largest channel, in which the current was very strong, and kept nearly parallel to, and about ten miles from, the Rein-deer mountains. Their outline, viewed from this distance, appeared very regular, the only remarkable parts being some eminences that were tinged with a deep pink colour. Sailing by one of the huts at a quick rate, every one's attention was arrested at hearing a shrill sound, which was supposed to be a human voice; but on landing to ascertain the fact, we could find no person, nor any footsteps. We, therefore, continued our journey. As we proceeded, the river became more devious in its course, the huts of the Esquimaux were now more frequent; none of them, however, seemed to have been recently inhabited. The islands were of the same alluvial kind as those seen yesterday, and the wood on them equally plentiful and large. We stopped to sup at nine, extinguished the fire as soon as we had finished, and then retired to sleep in the boat, keeping two men on guard.

Sunday, 14th. We set off aided by a fresh breeze this morning, and at the end of seven miles came to the last of the fir trees, in latitude 68° 40′ N., the only wood beyond this being stunted willows, which became still more dwarfish at thirty miles from the mouth of the river. There was plenty of drift-wood on the borders of the islands, and some even on the higher parts, at a distance from the water; from which it would appear that at certain seasons they are inundated. At length the main stream took a turn to the S.S.W., which we followed, though there was a branch northwards, but it seemed to be much impeded by mud-banks.* At the end of eight miles the river again inclined to the north of west, round the southern extremity of Halkett island, and there were openings to the north and south, which we did not stop to examine. A fog-bank hung over the northern horizon, which gave us no little uneasiness, from its strong resemblance to a continuous line of ice-blink; and the clouds, from the sun-beams falling on them, had the exact appearance of icebergs. However, the sun became sufficiently powerful in the afternoon to dissipate the cause of this illusion, and relieve us from anxiety on that score. A body of water, nearly equal to that we were descending, poured in between the Colville and Halkett islands with such force as to cause a very strong ripple at the point of junction, which we avoided by keeping close to the shore of Langley island. The channel, after the union of these streams, increased to a breadth of two miles, preserving a N.N.W. course. We stood twelve miles in this direction, and two to the westward, when we were gratified by the delightful prospect of the shore suddenly diverging, and a wide space of open water to the north-

* An attentive perusal of Sir Alexander Mackenzie's Narrative leads me to the conclusion, that it was this northern branch which that traveller pursued in his voyage to Whale Island.

ward, which we doubted not would prove to be the sea. Just at this time a seal made its appearance, and sported about the boat as if in confirmation of this opinion. We attempted to coast along the shore of Ellice island, but found the water too shallow, and that the boat grounded whenever we got out of the channel of the river, which was near the western side. The wind and waves were too high for us to make any progress in the middle of the stream, and as the clouds threatened more boisterous weather, we went to Pitt island to encamp. The haze which had hidden all distant objects since five P.M. passed off as the sun set, and we gained a very magnificent view of that portion of the rocky mountain which I have called after my companion Dr. Richardson, and of which the remarkable conical peak, named in honour of my friend Dr. Fitton, President of the Geological Society, and the Cupola mountain, are the most conspicuous objects. These were subsequently found to be near sixty miles distant. The water was entirely fresh, and there was no perceptible rise of tide. Our drowsy companion Baptiste, when he looked upon the vast expanse of water, for the first time, expressed some apprehension that we had passed the Forks, and that there was a doubt of our seeing the Indian chief; but he was by no means convinced of the fact until the following day, when he tasted salt water, and lost sight of the main shore. After our Sunday evening's supper, the party assembled in the tent to read prayers, and return thanks to the Almighty, for having thus far crowned our labours with success.

Monday, 15th.

In the morning of the fifteenth the wind blew a gale, as it had done through the night, and every object was obscured by a thick fog. About six A.M. we took advantage of a temporary abatement of the wind to cross over to some higher land on the eastern side, which we had seen the preceding evening, appearing like islands. Owing to the thickness of the fog, we were guided in our course at starting solely by the compass. When we reached the channel of the river, the gale returned with increased violence, and its direction being opposite to the current, such high waves were raised, that the boat took in a good deal of water. The fog now cleared away, and the three eminences mistaken for islands were ascertained to be conical hummocks, rising above the low eastern shore. We pushed for the nearest, and landed a short distance from its base at eight A. M. On going to the summit of this eminence, in the expectation of obtaining the bearings of several distant points, we were a little disappointed to find that only the low shores of Pitt Island were visible, extending from S.E. to W.N.W., though we were repaid for our visit by observing two moose deer quietly browsing on the tops of the willows, a short distance from us. Mr. Kendall hastened down to despatch Baptiste in pursuit of them, who returned an hour afterwards to inform us that he had wounded one, which he had been prevented from following by the loss of his powder-horn. As there was no possibility of our getting forward until the gale abated, Baptiste and Augustus were sent out to hunt, there being numerous tracks of moose and rein deer in the neighbourhood of the tent. I also despatched Mr. Kendall, with two seamen, to walk some distance into the interior, and endeavour to clear up the doubt whether we were upon the main shore, or upon an island. The astronomical observations obtained at the encampment place it in latitude 69° 3′ 45″ N., longitude 135° 44′ 57″ W. A tide-pole was put up immediately on our landing, and we perceived the water to rise about three inches in the course of the forenoon, and to fall the same quantity in the evening. The temperature of the air did not exceed forty-eight degrees all this day: when in the river, it used to vary from 55° to 70°. Mr. Kendall came back in the evening, bringing the agreeable intelligence that he had assisted in killing a female moose and her calf, and that Augustus had shot a rein deer. Some men were sent to carry the meat to the borders of a river which Mr. Kendall had discovered, while the boat went round to its entrance about one mile from the encampment. They returned at sunset. Many geese and ducks were seen by our hunters. Throughout the whole of Mr. Kendall's walk, of twelve or fourteen miles, he saw only the same kind of flat land, covered with the dwarf willow and the moose-berry plant, as was discovered from the tent, except one small lake, and the river that has been mentioned, issuing from it.

Tuesday, 16th. The atmosphere was so thick on the morning of the 16th as to confine our view to a few yards; we therefore remained at the encampment till the sun had sufficient power to remove the fog: temperature of the air 39°. Embarking at eleven A.M., we continued our course along the shore of Ellice Island, until we found its coast trending southward of east. There we landed, and were rejoiced at the sea-like appearance to the northward. This point is in latitude 69° 14′ N., longitude 135° 57′ W., and forms the north-eastern entrance to the main channel of the Mackenzie River, which, from Slave Lake to this point, is one thousand and forty-five miles according to our survey. An island was now discovered to the N.E., looking blue from its distance, towards which the boat was immediately directed. The water, which for the last eight miles had been very shallow, became gradually deeper, and of a more green colour, though still fresh, even when we had entirely lost sight of the eastern land. In the middle of the traverse, we were caught by a strong contrary wind, against which our crews cheerfully contended for five hours, though drenched by the spray, and even by the waves, which came into the boat. Unwilling to return without attaining the object of our search, when the strength of the rowers was nearly exhausted, as a last resource, the sails were set double-reefed, and our excellent boat mounted over the waves in the most buoyant manner. An opportune alteration of the wind enabled us, in the course of another hour, to fetch into smoother water, under the shelter of the island. We then pulled across a line of strong ripple which marked the termination of the fresh water, that on the seaward side being brackish; and in the further progress of three miles to the island, we had the indescribable pleasure of finding the water decidedly salt.

The sun was setting as the boat touched the beach, and we hastened to the most elevated part of the island, about two hundred and fifty feet high, to look around; and never was a prospect more gratifying than that which lay open to us. The Rocky mountains were seen from S.W. to W.½N.; and from the latter point, round by the north, the sea appeared in all its majesty, entirely free from ice, and without any visible obstruction to its navigation. Many seals, and black and white whales were sporting on its waves; and the whole scene was calculated to excite in our minds the most flattering expectations as to our own success, and that of our friends in the Hecla and the Fury. There were two groups of islands at no great distance; to the one bearing south-east I had the pleasure of affixing the name of my excellent friend and companion Mr. Kendall, and to that bearing north-east the name of Pelly was given, as a tribute justly due to the Governor of the Hudson Bay Company, for his earnest endeavours to promote the progress and welfare of the Expedition. A similar feeling towards my much esteemed friend Mr. Garry, the Deputy Governor of the Company, prompted me to appropriate his name to the island on which we stood,—a poor, indeed, but heartfelt expression of gratitude, for all his active kindness and indefatigable attention to the comfort of myself and my companions.

During our absence the men had pitched the tent on the beach, and I caused the silk union-flag to be hoisted, which my deeply-lamented wife had made and presented to me, as a parting gift, under the express injunction that it was not to be unfurled before the Expedition reached the sea. I will not attempt to describe my emotions as it expanded to the breeze —however natural, and, for the moment, irresistible, I felt that it was my duty to suppress them, and that I had no right, by an indulgence of my own sorrows, to cloud the animated countenances of my companions. Joining, therefore, with the best grace that I could command, in the general excitement, I endeavoured to return with corresponding cheerfulness, their warm congratulations on having thus planted the British flag on this remote island of the Polar Sea.

Some spirits, which had been saved for the occasion, were issued to the men; and with three fervent cheers they drank to the health of our beloved monarch, and to the continued success of our enterprize. Mr. Kendall and I had also reserved a little of our brandy, in order to celebrate this interesting event; but Baptiste, in his delight of beholding the sea, had set before us some salt water, which having been mixed with the brandy before the mistake was discovered, we were reluctantly obliged to forego the intended draught, and to use it in the more classical form of a libation poured on the ground.

Baptiste, on discovering that he had actually reached the ocean, stuck his feathers in his hat, and exultingly exclaimed, "Now that I am one of the *Gens de la mer*, you shall see how active I will be, and how I will crow over the *Gens du nord*," the name by which the Athabasca voyagers are designated. No fresh water was found on Garry Island until Augustus discovered a small lake, the streams that poured down from the cliffs being as salt as the sea. The temperature of the sea water was 51°; the fresh water we had left at five miles from the island 55°; and that of the air 52°.

Garry Island is about five miles long, by two broad, and seems to be a mass of frozen mud, which, in the parts exposed to the air and sun, has a black earthy appearance. It is terminated to the north-west by a steep cliff, through which protrude, in a highly inclined position, several layers of wood-coal, similar to that found in the Mackenzie. There was likewise observed a bituminous liquid trickling down in many parts, but particularly near the south-west point of the cliff where the bank had been broken away, and a hollow cavity was formed. The ravines and gullies were still filled with ice, though none was seen on the level ground. There were no stones above the sea level; those on the beach consisted of granite, greenstone, quartz, and lydianstone, of a small size and completely rounded. The vegetable productions were grasses, a few mosses, and some shrubs, the latter in flower. Four foxes were the only land animals we saw; and a small hawk, some gulls, dotterels, and phaleropes, composed the list of birds. A large medusa was found on the beach.

Wednesday, 17th.

The sky was cloudless on the morning of the 17th, which enabled us to ascertain the position of our encampment to be in latitude 69° 29′ N., longitude 135° 41′ W., and the variation of the magnetic needle to be 51° 42′ E. We likewise found that it was high water that day at one P.M. with a rise and fall of eight inches, but the direction of the flood could not be ascertained. I wrote for Captain Parry an account of our progress, with such information as he might require, in case he wished to communicate either with the Company's Post at Fort Good Hope, or our party, and deposited my letter, with many others that I had in charge for himself and the officers of the ships, under a pole erected for the purpose, on which we left a blue and red flag flying, to attract his attention. Another statement of our proceedings was encased in a water-proof box, and committed to the sea, a mile to the northward of the island. The wind blew strong off the land at the time, and there was a gale from the northwest the next day, so that there is every chance of the letter having made good way to the eastward.

Having completed the observations, we embarked at two P.M., and pulled along the western shore of the island three miles to the sandy spit at its south-west end, on which there was a vast quantity of drift-wood piled by the action of the waves. From this point we launched forth to cross towards the Mackenzie under double-reefed sails, as the wind was blowing strong, and the waves high in the offing; but finding the boat very stiff and buoyant, the sail was increased, and reaching the eastern point of Ellice Island by seven P.M. we encamped at the foot of the outermost of the three hummocks mentioned on the 15th of August. As we passed along the shore of the island, we disturbed some moose and rein deer, and several geese, cranes, and swans, that were quietly feeding near the water. At this period of the year, therefore, there would be no lack of food, in this country, for the skilful hunter. In the course of the evening I found that a piece of the wood-coal from Garry's Island, which I had placed in my pocket, had ignited spontaneously, and scorched the metal powder-horn by its side.

Our enterprising precursor, Sir Alexander Mackenzie, has been blamed for asserting that he had reached the sea, without having ascertained that the water was salt. He, in fact, clearly states that he never did reach the salt water. The danger to which his canoe was exposed in venturing two or three miles beyond Whale Island, (which lies to the eastward of our route,) at a time when the sea was covered with ice to the north, is a sufficient reason for his turning back; and we can abundantly testify that those frail vessels are totally unfitted to contend against such winds and seas as we experienced in advancing beyond the volume of fresh water poured out by the Mackenzie. It is probable, therefore, that even had the sea been free from ice at the time of his visit, he could not have gone far enough to prove its saltness, though the boundless horizon, the occurence of a tide, and the sight of porpoises and whales, naturally induced him to say that he had arrived at the ocean. The survey of the Mackenzie made on this Expedition, differs very little in its outline from that of its discoverer, whose general correctness we had often occasion to admire. We had, indeed, to alter the latitude and longitude of some of its points, which he most probably laid down from magnetic bearings only; and it is proper to remark, that in comparing our magnetic bearings with his, throughout the whole course of the river, they were found to be about fifteen degrees more easterly; which may, therefore, be considered as the amount of increase in variation since 1789. In justice to the memory of Mackenzie, I hope the custom of calling this the Great River, which is in general use among the traders and voyagers, will be discontinued, and that the name of its eminent discoverer may be universally adopted.

Thursday, 18th.

The excursions to Garry Island having made us acquainted with the state of the sea to the northward, and having shown that, the bank at the mouth of the river being passed, there was no visible impediment to a boat's proceeding eastward, I was desirous of making further examination in aid of the future operations of the Expedition, by going over to the western shore, and of reaching, if possible, the foot of the Rocky Mountains. With these intentions we embarked at nine A.M., but before we could get half way to the nearest part of Pitt Island, a gale of wind came on from N.W., followed by violent squalls, which, from the threatening appearance of the clouds, and the rapid descent of the thermometer from 68° to 51°, seemed likely to be of some continuance. The design was, therefore, abandoned, and the boat's head directed towards the entrance of the river. It proved, however, no easy task to get into the proper channel; and to effect this object the officers and crew had to drag the boat half a mile over a bar, while the waves were beating into it with such force as to make us apprehensive of its being swamped. As soon as we were in deep water, all the sail was set that the boat could bear, and at two P.M. we arrived at the narrow part. Here, likewise, the waves were high and breaking, and for the purpose of avoiding these and the strength of the current, we kept as close to the shore as possible, going through the water at seven miles an hour, and about four over the current. The wild fowl, warned by the sudden change of the weather, took advantage of this fair wind, and hastened away in large flights to the southward. At ten P.M., the boat having twice grounded, from our not being able to see our way clearly, we halted to sup, and laid down to sleep before a good fire. Temperature at 45°.

Friday, 19th.

When day-light permitted us to distinguish the channels, we embarked again, and scudded under the foresail before the gale, which this day blew with increased violence. We halted to breakfast near some winter habitations of the Esquimaux, which we supposed, from the freshness of the wood-shavings, and the implements of fishing that were scattered about them, had been abandoned only in the preceding spring; and as it was probable they would revisit this spot, we fixed to the pole of a tent a present of a kettle, knife, hatchet, file, ice-chissel some beads, and pieces of red and blue cloth. These huts were constructed of drift wood, in a similar manner to those which will be described in a subsequent part of the narrative. A second present was deposited at some other huts, and a third at those below the rapids. We imagined that some, if not all, of these would be found by the Esquimaux, and would make them acquainted with our visit. By noon we had advanced as far as the rapid, which we ascended under sail; and at a few miles above this point, owing to the fogginess of the atmosphere, we took a more western channel than that by which we descended. This proved circuitous, though it ultimately brought us to the former route. It was quite dark before we could find a secure place for the boat, and a sheltered spot for the tent. The gale continued without abatement, the weather was raw and cold, and it was with difficulty we collected some sticks to kindle a fire. Temperature 40°.

Saturday, 20th.

On the 20th the wind was moderate. We resumed our journey at four A.M.; past our sleeping-place of the 12th by noon, and at sunset encamped at the narrow part of the river where the numerous channels commence. Large flights of geese and swans were observed passing to the southward all this day. The musquitoes again made their appearance, though the temperature was at 45°: scarcely any of them had been seen on the descent to the sea.

Sunday, 21st. Temperature at day light, on the 21st, 37°. We commenced our labour under oars, but a strong gale from the southward soon rendered this mode of ascending the river ineffectual. The men were, therefore, divided into two parties, who towed the boat by line, relieving each other at intervals of an hour and a half. At fifty minutes past one P.M. we were abreast of the Red River, and there met a large party of the lower Loucheux Indians, who had assembled to wait our arrival. They welcomed our return with every demonstration of joy, more particularly that of Augustus and Baptiste, and at first cheerfully assisted the men in towing, but, like Indians in general, they soon became tired of this labour, and rather impeded than forwarded our progress. So we distributed to each a present; made known as well as well as we could by signs, that at our next visit we would purchase whatever fish or meat they might collect, and took our leave of them. Owing to the detention these men and another party occasioned, we were caught by a heavy gale from N.W. before we could reach our encampment at the head of the Narrows, and had to pitch the tent in pelting rain. Temperature 43°.

Monday, 22nd. On the 22nd, we started at four in a thick wet fog, which gave place to snow and sleet, and sailed the whole day before a strong N.W. wind, much to the annoyance of several Indians who tried to keep pace with the boat, by running along the shore: each of them had a present of tobacco thrown to him. We encamped near the bottom of Cannon-shot Reach; the weather was extremely cold, and, during the night, ice was formed in the kettle. On the next day the wind came contrary from S.E., which obliged us to have recourse to the tow-line. The frequent recurrence of sand-banks, to avoid which we had either to pull round or cross the river, made this day's operations very tedious. In turning round one of the points, we came suddenly upon a party of Indians, who had not seen us on our way down. Our appearance, therefore, created great alarm; the women and children were instantly despatched to the woods, and the men came down to the beach with their guns and arrows prepared, and knives drawn; but the explanation that Baptiste gave, soon allayed their fears. They were, indeed, objects of pity; all their property had been destroyed to testify their grief at the death of some of their relations, and the bodies of several were still sore from the deep gashes they

Tuesday, 23rd.

had inflicted on themselves in their demonstrations of sorrow. We distributed such useful articles among them as we had remaining, but the supply was not at all equal to their necessities. Several of them attempted to follow us in their canoes by poling, which they dexterously perform by pushing at the same time with a pole or paddle in each hand; the boat however, was towed faster than they could ascend the stream, and they were soon far behind. We arrived at six P.M. at the Trading River, and there met another party of the Loucheux, among whom was the woman whose tears had excited our sympathy on the 11th, now in high glee, and one of the most importunate for beads. The boy was likewise there to whom the tea had been given as a remedy for his fever, completely recovered, which was, no doubt, ascribed to the efficacy of the medicine. Not choosing to encamp near these people, we crossed the river, and towed four hours longer, when we reached Fort Good Hope. Mr. Dease, and all his fort, were overjoyed on seeing us again, because the Indians had begun to surmise, and in fact had brought a report that we had all been massacred by the Esquimaux; and had we been detained another week, this statement would have gained entire credence, and, in all probability, spread throughout the country.

The Indian whose fish we carried on our way down, happened to be at the fort, and he cheerfully communicated, through the interpreter, a female, all the information that he or his tribe possessed respecting the mouth of the river, the sea-coast, and the Esquimaux, all topics highly interesting to us, but we subsequently found that his knowledge of these matters was very imperfect. We made known to him our wish that the Esquimaux should be informed of our arrival as soon as possible, and signified that a very substantial present would be given to any person that would carry the intelligence to them in the course of the following winter. Mr. Dease pressed this point strongly on his consideration. This gentleman, indeed, was anxious to promote our desires in every respect, and promised that his utmost exertions should be used to procure a good supply of provision for our next summer's voyage, though he represented the hunters in this vicinity as unskilful and inactive, and begged of me not to rely too much on his collection. We left in his charge five bags of pemmican, and the superfluous stores, to lighten the boats. We quitted the fort in the afternoon with a contrary wind, and towed twenty miles up the stream before we encamped, though the beach was composed of sharp stones, which rendered walking very unpleasant.

The wind being contrary during the four following days, we could only ascend the river by using the tracking line. Our crew cheerfully performed this tedious service, though three of them had been much reduced by dysentery, brought on by previous fatigue, exposure to wet, and by their having lived for some time on dried provision. These men, however, had gradually been gaining strength since the fresh meat was procured on Ellice Island.

On the 25th we came to the aspen, poplar, and larch, in latitude 67° 10′ N., and were not a little surprised to observe the change in their foliage within the last fortnight. Their leaves had assumed the autumnal tint, and were now fast falling. The wild fowl were hastening in large flocks to the south, and every appearance warned us that the fine season drew near its close.

28th. In the passage through the rampart defile, several families of the Hare Indians were observed encamped on the heights, for the purpose of gathering berries which were at this time ripe, and in the best flavour. At the first sight of the boat the women and children scampered down wherever descent was practicable, to get at their canoes, that they might cross over to us, but we travelled so fast that only a few could overtake the boat. The Indians who reside near this river, from their want of skill in hunting, principally subsist, from spring to autumn, on the produce of their fishing nets, and on wild berries. At the influx of small streams, or wherever there is any eddy, a net is set. In shallow water it is suspended upon sticks planted in a semicircle, so as to enclose the mouth of the river, or the sweep of the eddy; but where the water is deep, and the shore bold or rocky, two stout poles are firmly secured at a short distance from the water's edge, the breadth of a net apart, to the ends of which pliable rods are fastened, of a length sufficient to hang over the water, and to these the net is attached. In the winter these Indians snare hares, which are very abundant in this quarter.

29th. On the 29th we arrived at the upper rapids, which were scarcely discernible at the time of our descent; but from the falling of the water since that time, there was a dry sand-bank of considerable extent in the centre, and the waters on each side of it were broken and covered with foam. Augustus being tired with tracking, had wandered from us to the extremity of this bank, from whence he could not be extricated without great hazard, unless by making him return to the bottom of the rapid. As this, however, would have compelled the poor fellow to pass the night upon the sand-bank, Mr. Kendall undertook to bring him off, by running with the current to the point at the commencement of the rapids, which he effected in a masterly manner, although the boat struck twice, and was in considerable danger from the violence of the eddies.

We found, at the place of our encampment, a solitary old woman, sitting by a small fire, who seemed somewhat alarmed at her visitors, until she was joined, after dark, by her husband and son. As soon as the man understood from our signs that we were desirous of having some fish for supper, he instantly embarked to examine his nets; but as they proved to be empty, the woman generously dragged a pike out of a bundle on which she was sitting, and presented it to us, though it was evidently reserved for their own meal. In return we furnished them with a more substantial supper, and made them some useful presents. The weather was extremely sultry throughout this day; at two P.M. the thermometer stood in the shade at 66°, and at 76° when exposed to the sun. The refraction of the atmosphere, which we had often remarked to be unusually great since we had entered the Mackenzie, was this day particularly powerful. The mountains were distorted into the most extraordinary shapes, and the banks of the river, which we knew to be only from thirty to sixty feet high, appeared to have such an elevation, that it would have been impossible for us to recognise the land. The air became cooler in the evening, and the atmosphere less refractive. Soon after sunset the objects appeared in their proper form, and we enjoyed the prospect of the delightful mountain scenery that distinguishes this rapid.

Tuesday, 30th.

Favoured by a N.W. gale, we made great progress on the 30th. The temperature of the air varied in the course of the day from 62° to 41°. The brulôts and sand-flies were very teazing wherever we landed; but these, unlike the musquitoes, disappear with the sun.

The upper parts of the Rocky Mountains on the western side of the river were, at this time, covered with snow, but not those of the eastern side, which are, probably, less elevated than the former. We had no opportunity of ascertaining their height, though we conjectured that the loftiest did not exceed two thousand feet, as it was free from snow in the early part of August.

September, 1st.

At sunset this evening we quitted the muddy waters of the Mackenzie, and entered the clear stream that flows from the Great Bear Lake; but owing to the shallowness of the water near its mouth, and the beach being a mere collection of stones, we had to grope our way long after dark in search of a place for an encampment, stumbling and falling at every step. At length we espied a light about a mile further up the river on the opposite shore; we, therefore, crossed over, at the expense of some heavy blows to the boat, and tracked along the base of a steep bank, until we reached the fire. There we found a Canadian and two Indian boys who had been sent from Bear Lake three days before in a canoe, to procure some white mud from the banks of the Mackenzie to decorate our houses. This man was the bearer of a letter from Lieutenant Back to me, which detailed the proceedings at the Fort.

Friday, 2nd. We embarked at daylight, having the canoe in company. The weather was cold and raw throughout the day; the temperature from 34° to 45°; but the party were kept in constant exercise, either in tracking or walking; the steersman and bowman only being required in the boat. Except where the river was bounded by steep cliffs, the path was pretty good. Its general breadth varied from three hundred to five hundred yards, and its banks were tolerably well wooded, but the trees were small.

Saturday, 3rd. This morning the ground was firmly frozen, and the thermometer stood at 28°, when we commenced our operations. Early in the afternoon we arrived at the lower part of the mountain, and which we had kept in view this day, and the greater part of the preceding. As we had now to ascend a succession of rapids for fifteen miles, and two of our crew were lame, I directed the canoe to be laid up on the shore, and took the Canadian and the boys to assist at the tow-line. We had not advanced more than two miles before we met with an accident that was likely to have been attended with serious consequences: in the act of hauling round a projecting point, and in the strength of the current, the tow-line broke, and the boat was driven with great force against a large stone at some distance from the shore, having deep water on every side. There it lay with the broadside exposed to the whole pressure of the current, beating violently against the stone; and from this situation it could not have been extricated, had not Gustavus Aird, the strongest man of the party, ventured to wade into the river, at the imminent risque of being swept off his feet, until he could catch the rope that was thrown to him from the boat. As soon as it was dragged to the shore, we found that part of the keel was gone, and the remainder much twisted, and all the fastenings of the lowest plank were loosened. The carpenter set to work to repair this mischief in the best manner he could with the materials he had, and before night the boat was again launched. The leaks, however, could not be quite stopped, and in our further progress one of the men was constantly employed baling out the water.

Sunday, 4th.

The next day's operations were tedious and hazardous as long as the rapids continued. The men had to walk with the tow-line along a narrow ledge that jutted out from the base of a steep rocky cliff, which was very slippery from the rain that had fallen in the night: a false step might have proved fatal; and we rejoiced when, having passed the rapids, we found earthy banks and a better path. The services of Augustus and the Indian lads being no longer required, I despatched them to the Fort, to apprize the party there of our approach.

We had a severe frost this night: at daylight in the morning the thermometer was down to 20°, and a raw fog contributed to make the weather very cold and comfortless. The sun shone forth about eleven, and soon dispersed the fog, and then the temperature gradually rose to 54°.

Monday, 5th.

At four P.M. we arrived at the foot of the upper rapid, and in two hours afterwards entered the Great Bear Lake, and reached the house at seven. Dr. Richardson having returned from his voyage to the northern part of the lake, the members of the Expedition were now, for the first time, all assembled. We heartily congratulated each other on this circumstance, and also on the prospect of being snugly settled in our winter-quarters before the severe weather. Dr. Richardson had surveyed the Bear Lake to the influx of Dease's River, near its N.E. termination, at which point it is nearest to the Coppermine River. He fixed upon the first rapid in Dease River as the best point to which the eastern detachment of the Expedition could direct its steps, on its return from the mouth of the Coppermine River the following season. The rapid was, by observation, in latitude 66° 53′ N., and longitude 118° 35′ W., and the variation of the magnetic needle there, was 47° 29′ E.

THE FOLLOWING TABLE CONTAINS THE DISTANCES TRAVELLED BY THE EXPEDITION DURING THE SUMMER OF 1825.

Principal Places.	Statute Miles.
From New York to Penetanguishene, by the route we travelled	760
Lake Huron - - - - - - - - -	250
Lake Superior - - - - - - - - -	406
From Fort William to Cumberland House - - - - -	1018
Cumberland House to Fort Chipewyan - - -	840
Chipewyan to Fort Resolution, Slave Lake - - -	240
Fort Resolution to the commencement of the Mackenzie	135
Head of the Mackenzie to Fort Simpson - - -	103
Fort Simpson to Bear Lake River - - - -	271
Bear Lake River to, and the return from, Garry Island	1206
Length of the Bear Lake River to the Fort - - - -	91
Dr. Richardson's excursion to the north-east termination of Bear Lake - - - - - - - - - -	483
Distance travelled - - - -	5803
Number of Miles surveyed - -	2593

CHAPTER II.

TRANSACTIONS AT FORT FRANKLIN, 1825-26.

Mr. Dease having passed the winter of 1824-25 at the Big Island of Mackenzie, arrived here with fifteen Canadian voyagers, Beaulieu, the interpreter, and four Chipewyan hunters, on the twenty-seventh of July, 1825; which, on account of the drifting of the ice, was as soon as he could, with safety, ascend the Bear Lake River. Several of the Dog-Rib Indians were on the spot, which enabled him to take immediate steps towards procuring a supply of dried meat for our winter use, as well as of fresh meat for present consumption. It having been ascertained that the Rein-deer are most abundant in the north-east quarter of the lake, during the months of August and September, a select party of Indians was despatched to hunt thereabout, under the direction of the interpreter, who took a large canoe for the purpose of bringing home the produce of their hunt. Other men were sent to inform the Hare Indians of our wish to purchase any meat they might bring to the establishment. Our principal subsistence, however, was to be derived from the water, and Mr. Dease was determined in the selection of the spot on which our residence was to be erected, by its proximity to that part of the lake where the fish had usually been abundant. The place decided upon was the site of an old fort belonging to the North-West Company, which had been abandoned many years; our buildings being required of a much larger size, we derived very little benefit from its materials. The wood in the immediate vicinity having been all cut down for fuel by the former residents, the party was obliged to convey the requisite timber in rafts from a considerable distance, which, of course, occasioned trouble and delay. We found, however, on our arrival, all the buildings in a habitable state, but wanting many internal arrangements to fit them for a comfortable winter residence. They were disposed so as to form three sides of a square, the officers' house being in the centre, those for the men on the right, with a house for the interpreter's family, and the store on the left. A blacksmith's shop and meat store were added, and the whole was inclosed by the stockading of the original fort, which we found highly serviceable in skreening us from the snow-drift and wintry blasts. The officers' dwelling measured forty-four feet by twenty-four, and contained a hall and four apartments, beside a kitchen. That of the men was thirty-six feet by twenty-three, and was divided into three rooms. These buildings were placed on a dry sandy bank, about eighty yards from the lake, and twenty-five feet above it; at the distance of a half a mile in our rear, the ground rose to the height of one hundred and fifty feet, and continued in an even ridge, on which, though the timber had been felled, we found plenty of small trees for fuel. This ridge bounded our view to the north; and to the west, though confined to less than two miles, the prospect was pretty, from its embracing a small lake, and the mouth of a narrow stream that flowed in at its head. Our southern view commanded the south-west arm of Bear Lake, which is here four miles wide, and not deeper than from three to five fathoms, except in the channel of the river, which conveys its waters to the Mackenzie. We had also, in front, the Clark-hill, a mountain about thirty-six miles distant, which was always visible in clear weather. When the refraction was great, we saw the tops of some other hills, belonging to the range that extends from Clark-hill to the rapid in Bear Lake River.

Immediately under the sandy soil on which the house stood, there is a bed of tenacious bluish clay, of unknown thickness, which, even in the months of August and September, was firmly frozen at the depth of twenty-one inches from the surface. No rocks were exposed in any part, and wherever the surface had been torn up, a clayey soil appeared. Many boulder stones of granite, limestone, sandstone and trap rocks, were scattered about the lake, not far from the shore.

The trees at some distance from our fort consisted of black and white spruce, and larch, generally small, though a few of the better grown measured from four to five feet in girth, and were from fifty to fifty-five feet high. Dr. Richardson ascertained, by counting the annual rings, that some of them, in a sound state, were upwards of one hundred and thirty years old; while others, which were not much greater in size, had two hundred and fifty rings, but these were decayed at the heart.

The officers had done me the honour, previous to my arrival, of giving the name of Franklin to the fort, which I felt a grateful pleasure in retaining at their desire, though I had intended naming it Fort Reliance. The number of persons belonging to the establishment amounted to fifty: consisting of five officers, including Mr. Dease; nineteen British seamen, marines, and voyagers; nine Canadians; two Esquimaux; Beaulieu, and four Chipewyan hunters; three women, six children, and one Indian lad; besides a few infirm Indians, who required temporary support. This party was far too large to gain subsistence by fishing at one station only; two houses were, therefore, constructed at four and seven miles distance from the fort, to which parties were sent, provided with the necessary fishing implements; and not more than thirty persons were left to reside at the principal establishment. From fifteen to twenty nets were kept in use, under the superintendence of Pascal Coté, an experienced fisherman, who had two assistants. These were placed opposite the house, and towards the end of summer, and in autumn, they yielded daily from three to eight hundred fish, of the kind called "the Herring Salmon of Bear Lake," and occasionally some trout, tittameg, and carp. Four Dog-Rib Indians, who were engaged to hunt the Rein-deer in the neighbourhood of the fort, from want of skill, contributed very little fresh meat to our store. Augustus and Ooligbuck employed themselves in the same service, but from not being accustomed to hunt in a woody country, they were not more successful.

The consideration of next importance to furnishing the party with food, was to provide regular occupation for the men, who had not the resources to employ their time which the officers possessed. Accordingly, some were appointed to attend exclusively to the fishing nets, others to bring home the meat whenever the hunters killed any deer; some were stationed to fell wood for fuel, others to convey it to the house, and a third set to split it for use. Two of the most expert travellers on snow-shoes were kept in nearly constant employment conveying letters to and from the posts in the Mackenzie and Slave Lake. As the days shortened, it was necessary to find employment during the long evenings, for those resident at the house, and a school was, therefore, established on three nights of the week, from seven o'clock to nine, for their instruction, in reading, writing, and arithmetic; and it was attended by most of the British party. They were divided in equal portions amongst the officers, whose labour was amply repaid by the advancement their pupils made: some of those who began with the alphabet, learned to read and write with tolerable correctness. Sunday was a day of rest; and, with the exception of two or three of the Canadians, the whole party uniformly attended Divine service, morning and evening. If, on the other evenings for which no particular occupation was appointed, the men felt the time tedious, or if they expressed a wish to vary their employments, the hall was at their service, to play any game they might choose; and on these occasions they were invariably joined by the officers. By thus participating in their amusements, the men became more attached to us, at the same time that we contributed to their health and cheerfulness. The hearts and feelings of the whole party were united into one common desire to make the time pass as agreeably as possible to each other, until the return of spring should enable them to resume the great object of the Expedition.

The officers found employment in making and registering the thermometrical, magnetical, and atmospherical observations, which were hourly noted from eight A.M. to midnight; and, in addition to the duties which they had in common, each had a peculiar department allotted to him.

Lieutenant Back had the superintendence of the men; and the accurate drawings which he finished during the winter, from sketches taken on the voyage, afford ample proof of his diligence and skill. Dr. Richardson, besides the duties of medical officer, which, from the numerous applications made by the natives, were not inconsiderable, devoted his attention to natural history, as well as to a series of observations on the force of the sun's radiation. Mr. Kendall constructed all the charts after the data had been recalculated by myself; he also made several drawings; and he undertook an interesting series of observations on the velocity of sound. To Mr. Dease the charge was committed of whatever related to the procuring and issuing of provision, and the entire management of the Canadian voyagers and Indians.

Previous to the officers leaving London, Dr. Fitton, President of the Geological Society, had the kindness to devote much of his time to their instruction in geology; and having furnished them with a portable collection for the purpose of reference on the voyage, Dr. Richardson, when he had leisure, explained these specimens, weekly, to the party, and assisted them in reading on this science, which proved a most agreeable and useful recreation to us all.

Some of the preceding remarks refer to a period of our residence later than that which I am about to enter upon; but I thought it best to insert them here, that the mention of them might not interrupt the narrative of occurrences which I shall now resume.

Thursday, 8th. On September 8th, two men were sent off to Slave Lake, in a canoe, with a despatch, containing an account of our proceedings, addressed to His Majesty's

Secretary of State for the Colonies; and as we expected letters from England, by the way of Hudson's Bay, they were directed to await their arrival at Slave Lake. There was almost constant rain from the 11th to the 14th, which much retarded the work going on out of doors, and particularly the construction of an observatory, which we were desirous of completing as soon as possible, that the magnetical observations might be commenced. We found employment, however, in whitewashing and fitting up the interior of the different houses. The 15th proving fine, we established a meridian line, and ascertained the variation by each of the compasses.

Tuesday, 20th.

Beaulieu returned with his family, the Chipewyan hunters, and some Dog-Ribs, bringing a supply of dried meat, rein-deer tongues, and fat, sufficient for a month's consumption, which was reserved for use when the fishing should become unproductive. These men reported, that at the time they quitted the northern shores of the lake, the deer were retiring towards this quarter; which intelligence accounted for the Indians having killed four within a day's march from the house.

Friday, 23rd.

The chimney of the last of the buildings being completed this morning, the flag-staff erected, and all the men assembled, we commemorated these events by the festivities usual on the opening of a new establishment in this country. The first part of the ceremony was to salute the flag; the men having drawn themselves up in line, and the women and children, and all the Indians resident at the fort, being disposed in groups by their side, a deputation came to solicit the presence of the officers. When we appeared, we found our guns ornamented with blue ribbons, and we were requested to advance and fire at a piece of money which was fastened to the flag-staff. The men then fired two volleys and gave three hearty cheers, after which Wilson the piper struck up a lively tune, and placing himself at the head of his companions, marched with them round to the entrance of the hall, where they drank to His Majesty's health, and to the success of the Expedition. In the evening the hall was opened for a dance, which was attended by the whole party, dressed in their gayest attire. The dancing was kept up with spirit to the music of the violin and bag-pipes, until day-light.

Monday, 26th.

These entertainments over, Beaulieu and the hunters were despatched to the chace, and they soon added two moose-deer to our store.

Tuesday, 27th.

There had been much rain in the course of the preceding week, and the temperature was generally mild, but a fall of snow took place on the 27th. Some Dog-Ribs came to the fort on that day with the produce of their autumnal hunt, which was very inconsiderable, but they rendered good service to us by taking away with them several of their relations, who had been subsisting on our bounty for some time. After their departure there only remained one man of the tribe, who, being afflicted with rheumatic fever, was retained under the care of Dr. Richardson. Warm clothing was provided for him, and a comfortable leathern lodge was erected for himself and family.

October, 1st. The month of October commenced with frost and snow, and the party were now furnished with fur caps, leathern mittens and trowsers, and the rest of their warm winter-clothing. This day we completed the erection of the observatory, and adjusted an instrument to the magnetic meridian, for the purpose of observing the variations of the needle.

Tuesday, 11th. Much snow fell on the night of the 7th, and on the 11th the small lake was firmly frozen over, and the ground in the same state. All the migratory birds being now gone, except a few ducks, which still lingered in the open water of Bear Lake, we considered this day to be the first of the winter. It was remarkably clear and fine, and we hailed the commencement of this season with a degree of pleasure, from its contrast with the wet unsettled weather which marks the close of summer. A few clouds passing over the sun's disk, produced an instantaneous depression of ten degrees of the mercury in a thermometer exposed to the sun's rays. The atmospherical refraction was remarkably strong at this time. We had repeated opportunities, in the course of the winter, of observing it to be greatest in similar states of the atmosphere.

The boats were now secured for the winter in a sheltered place, and screened as much as possible from the effects of the wind and snow drift, by a strong fence made of boughs and branches.

Friday, 14th. We were surprised on the 14th by the arrival of two Canadians from Fort Norman, with letters from Governor Simpson, and other gentlemen in the southern districts, containing satisfactory answers to the requisitions for stores that I had made in my passage through the country. We were also pleased to learn that Thomas Matthews, the carpenter, whom we had left at Cumberland House, on account of his leg being broken, had reached Fort Norman, in the Company's canoe; and I felt much indebted to Mr. James

Keith, and Mr. Smith, Chief Factors, for the care and tenderness with which they had conveyed him through the country.

The season at which the ice begins to form, is the most favourable for fishing in the lakes of this country, and we then procured from four to five hundred daily. Those not required for immediate consumption, were hung on a stage to freeze, in which state they keep until the following spring. But we could not derive the full advantage from the season, because the drift ice, making it unsafe to keep the nets set in Bear Lake, they were taken up on the 18th. Near a month elapsed before they could be set with safety under the ice; our first attempts resulting in the loss of three nets. We procured, however, a few fish from the small lake, during this interval, and the rest of our food was supplied from the store of dried meat.

Thursday, 20th.

We were visited on the 20th by a storm of snow, which continued, without intermission, for thirty-six hours. Although it put an end to the skating, and the games on the ice, which had been our evenings' amusement for the preceding week, yet the change made every one glad, because the snow was now deep enough for winter travelling. We had learned, some days before, that the hunters had stored fifteen rein-deer in the woods, and on the 22nd four men were despatched with sledges to bring them to the fort.

The first throw off of the dog-sledges for the season never fails to attract general attention; accordingly the whole party was collected to witness it on this occasion. They set off at full speed, and were soon out of sight. From this time dog-sledges were used to drag the fuel, which had been hitherto done by the men. We sent a party to cut down timber, and saw it into planks, fit for the construction of another boat.

Wednesday, 26th.

On the 26th the thermometer first fell below zero, but the month closed with a very calm, mild day. Mr. Kendall and I were employed in measuring a geographical mile on the small lake, preparatory to a series of observations on the velocity of sound. The only ferine companions we now had were a few hardy quadrupeds and birds, capable of enduring the winter. The variety of the former was confined to wolves, foxes, martens, hares, mice, and a few rein-deer. Of the feathered tribe, there were the raven and Canadian crow, some snow-birds, wood-peckers, red-caps, crossbeaks, Canada, rock, and willow partridges, and a few hawks and owls.

Having received information that the Hudson's Bay Company intended sending their annual despatch from the Mackenzie River to York Factory, by the close of this month, and the ice on Bear Lake and the Mackenzie River being, on the 9th, sufficiently strong, we forwarded a packet of letters to Fort Norman, and a dog-sledge to convey Thomas Matthews to this place. On the 15th the nets were reset under the ice, and we were relieved from the necessity of putting the party on short allowance. We had the additional pleasure of learning that the hunters had killed ten rein-deer. The men returned from Fort Norman on the 18th, accompanied by Thomas Matthews, whose leg was yet too weak for him to walk more than a short distance.

November, 9th.

During the middle, and towards the close of November, parheliæ were frequent; the most brilliant appeared on the 27th; it continued as long as the sun was above the horizon. The atmosphere was cloudless, and apparently free from haze, except just about the sun, which seemed to gleam through a fog. The surrounding circle was nearly complete, and displayed the prismatic colours vividly; from the centre of the sun's disk a beam of bright light extended upwards several degrees beyond the circle. The inner radius of the circle measured 21° 34′, and the outer 22° 50′. The wind blew fresh all the day from E.N.E., and the temperature was 10°. In the evening the moon was encircled by two distinct halos; temperature 7°.

Tuesday, 29th.

This morning the principal leader of the Dog-Ribs, and a large party of his tribe, came to the Fort. It is usual for Indians, on the first visit to an establishment, to make their approach in line, with much formality; but on this occasion our visitors showed an unusual degree of caution. Their distrust had originated in a very trifling occurrence at the close of our house-warming festivities on the 23rd of September. Some of the Canadians having asked Mr. Dease if our Highlandmen did not come from the same country with the rest of the English party, were told that they were natives of the mountainous lands, or *Montagnards*. This name unfortunately being used by the voyagers to designate the Dog-Ribs, was considered by the Highlanders to be a term of reproach when applied to themselves, and a scuffle ensued. Harmony was soon restored by the officers sending the most noisy to bed, and next morning the true meaning of the word Montagnard was explained to the Highlandmen, and the party set about their usual occupations with their wonted good feeling towards each other. Not so with an unlucky Dog-Rib, who had been attracted to the scene by hearing the name applied by the voyagers to his countrymen bandied about from one to the other, and thrusting his head into the crowd had received a blow. This at once confirmed all his fears, and he fled to spread a report amongst his countrymen that the white people intended to destroy the Indians. Although his report was not fully believed, yet it produced the feeling of distrust which the Indians manifested on their approach to the house. It was entirely removed by the explanation we gave. These Indians having brought a quantity of furs for the Hudson's Bay Company, as well as dried meat for ourselves, and I having understood from Mr. Dease that it would be an accommodation to them if they were permitted to deposit their furs at this place, instead of carrying them to Fort Norman, I acceded to this suggestion, and directed Mr. Dease to advance from our stores the goods required for the purchase of the furs, which were to be returned when we should visit that fort in the spring.

An old man belonging to the Company's establishment at Fort Norman arrived this day with his wife, to stay some time with us, because the supply of provision had failed at that post. We felt much pleasure in sharing our means with this aged couple, who were much reduced by their late scanty fare.

The close of November was marked by a succession of strong east winds, and a mildness of temperature, rare at this season. On the 30th the thermometer rose from + 18° to 29° above zero, on the occurrence of a gale from the north.

December, 1st.

The first of December being a cloudless day, we endeavoured to observe the latitude at noon, but failed, owing to the extraordinary atmospherical refraction.

Friday, 2nd.

At midnight, on the 2nd, there was a shower of hail, so small that we could hardly distinguish it from rain. Dr. Richardson thought he perceived lightning. Temperature + 22°, calm. On the night of the 4th another instance of a sudden increase of temperature from + 7° to 26° was observed, on a north wind succeeding a calm.

The fishery having gradually declined for some days, our nets were removed nearer to the entrance of Bear Lake River, where the current continued to keep the water open for a considerable space. We then procured a daily supply of fish sufficient for the rations of the household, as well as the dogs, though our number was now increased by the party from the more distant fishery, which had proved unproductive. The allowance was seven of the herring salmon to a man per day, and two to each dog.

The shortness of the days now precluding the Indians from

hunting, many came, according to their custom, to spear fish at the head of Bear Lake River, and their numbers gradually increased. They were not, however, successful, nor diligent, preferring to beg what they could from us, and sending their women and children to subsist on the offal of the fish used at the fort. To encourage them to greater exertion, I provided them with nets, and other fishing materials, but their indolence led them to make a very ungrateful return; for on several occasions they emptied our nets in the night, and thus not only robbed us of what they took away, but, by deranging the nets, deprived us of the whole of that day's supply. We never could ascertain the perpetrators of these thefts. The blame was invariably thrown on some aged and infirm men, who denied it. Notwithstanding the straits to which they became reduced, they could not be persuaded to go off to a more productive fishery, until we were compelled to withhold all supplies, from fear of starving our own party. These Indians showed more indolence, and less regard for truth and honesty, than any other tribes with which we had dealings. Their sufferings are often extreme, and some of them perish every year from famine; although, from the abundance of fish in this country, but slight exertion would be required to lay up, at the proper seasons, a stock for the whole year.

The difficulty of procuring nourishment frequently induces the women of this tribe to destroy their female children. Two pregnant women of the party then at the fort, made known their intention of acting on this inhuman custom, though Mr. Dease threatened them with our heaviest displeasure if they put it into execution: we learned that, after they left us, one actually did destroy her child; the infant of the other woman proved to be a boy. Infanticide is mentioned by Hearne as a common crime amongst the northern Indians, but this was the first instance that came under our notice, and I understand it is now very rare amongst the Chepewyan tribes;—an improvement in their moral character which may be fairly attributed to the influence of the traders resident among them.

Sunday, 18th. On the 18th a party of sixteen Hare Indians, two Copper Indians, and a Loucheux, arrived with sledges of dried rein-deer meat and furs. While the house was in confusion from the unpacking of their lading, a melancholy scene took place, which excited the warmest sympathy. The wife of one of our Dog-Rib hunters brought her only child, a female, for medical advice. As she entered the room it was evident that the hand of death was upon it. In the absence of Dr. Richardson, who happened to be out, all the remedies were applied that were judged likely to be of service; and as soon as he returned, there being yet a faint pulsation, other means were tried, but in vain. So gentle was its last sigh, that the mother was not at first aware of its death, and continued to press the child against her bosom. As soon, however, as she perceived that life had fled, she cast herself on the floor in agony, heightened by the consciousness of having delayed to seek relief till too late, and by apprehension of the anger of her husband, who was doatingly attached to the child. The Indians evinced their participation in her affliction by silence, and a strong expression of pity in their countenances. At the dawn of day the poor creature, though almost exhausted by her ceaseless lamentation, carried the body across the lake for interment.

Tuesday, 20th.

The 20th being a very stormy day, we were surprised at the arrival of two voyagers from Fort Good Hope, bearers of letters from Mr. C. Dease, conveying the gratifying intelligence that the Loucheux had seen the Esquimaux since the autumn, and that the latter had found the presents which had been left at their huts, and would be delighted to welcome the return of the white people to the Esquimaux lands next spring.

Thursday, 22d.

Our constant occupations had made the time pass so swiftly, that the shortest day came almost unexpectedly upon us. The sun rose this morning, (the 22d,) at $10^h\ 24^m$, thirteen minutes earlier than its appearance was expected from calculation, owing to the great refraction. Mr. Kendall and I measured its meridional altitude from the lake with two instruments, the one bringing its upper limb to the top of the land four miles distant, the elevation of which had been ascertained to be eight minutes, and the other to its base, the depression of which was two minutes. The mean of both these observations, corrected for refraction by the tables in the Nautical Almanack, gave a result of 65° 11′ 56″ N., which latitude exactly corresponds with the best observations made in the preceding autumn. At $8^h\ 30^m$ P.M. a halo was observed, whose radius measured 28° 40′ from the moon; and at an equal altitude with the latter body there were two paraselenæ, which, as well as the moon, were intersected by a luminous circle, having the zenith for its centre, and a diameter of 94° 15′. The length of our shortest day did not exceed five hours, but the long nights were enlivened by most brilliant moon-light, and we had frequent and very fine appearances of the Aurora Borealis. The latter phenomenon made some of its grandest displays on the 26th of October, the 2d of November, and the 7th of December. On all these occasions the disturbed motions of the magnetic needle were very remarkable, and a most careful series of observations convinced the party that they had a close connexion with the direction of the beams of light of which the aurora was composed. My observations also led me to conclude that the deviations of the needle were, in a certain degree, connected with changes in the weather; for, previous to a gale or a snow-storm, the deviations were always considerable; but during the continuance of the gale, the needle almost invariable remained stationary.

Preparations were made for the celebration of Christmas. The house was replastered with mud, all the rooms white-washed and repainted, and Matthews displayed his taste by ornamenting a chandelier with cut paper, and trinkets. On the evening of the 24th the Indian hunters' women and children were invited to share in a game of snap-dragon, to them an entire novelty. It would be as difficult to describe the delight which the sport afforded them after they recovered their first surprise, as to convey the full effect of the scene. When the candles were extinguished, the blue flame of the burning spirits shone on the rude features of our native companions, in whose countenances were pourtrayed the eager desire of possessing the fruit, and the fear of the penalty. Christmas Day falling on a Sunday, the party were regaled with the best fare our stores could supply; and on the following evening a dance was given, at which were present sixty persons, including the Indians, who sat as spectators of the merry scene. Seldom, perhaps, in such a confined space as our hall, or in the same number of persons, was there greater variety of character, or greater confusion of tongues. The party consisted of Englishmen, Highlanders, (who mostly conversed with each other in Gaelic,) Canadians, Esquimaux, Chipewyans, Dog-Ribs, Hare Indians, Cree women and children, mingled together in perfect harmony. The amusements were varied by English, Gaelic, and French songs. After these holidays were over, the Dog-Ribs at length yielded to the repeated solicitations of Mr. Dease, and removed in a body to a distant part of the lake, where they now confessed the fishery was more abundant. As the hunters were drawing rations from our store, he despatched them in quest of deer, furnishing them also with nets. After which there remained at the establishment, only one infirm Indian and his wife.

January, 1st.

January 1st, 1826. This morning the men called in the hall to offer the congratulations of the season to the officers, and we afterwards assembled to read divine service. On the evening of the 2nd, similar festivities were held to those at Christmas, to welcome the new year. The temperature was at —49° on the 1st, which was its lowest state during this winter. This severe weather was of short continuance, for on the 3rd there was a storm of snow, and the thermometer rose to —9°.

Accompanied by Mr. Dease, and Fuller, the carpenter, I walked several miles in search of birch-trees fit for the keel and timbers of the new boat. We found some that would answer for the latter purpose, but none for the keel; we, therefore, substituted pine. The general depth of snow in the woody and sheltered parts was two feet. On the 16th, by the return of the two men who had been sent to Slave Lake, we had the happiness of receiving a packet of letters, which left England in the preceding June. Beside the more interesting private communications, our friends had been kind enough to forward piles of newspapers, and several periodical publications. The 'Quarterly Review,' the 'Edinburgh Philosophical Journal,' and a series of the 'Literary Gazette,' and the 'Mechanics' Magazine,' were spread upon the table, and afforded us the most agreeable amusement, as well as never-failing topics for conversation. Could any of our friends have dropped in upon us, in the evening, they would have found us discussing the events of the by-gone year, with all the earnestness and interest which we could have shown had they been the occurrences of the day, and depended upon our decision. This valuable packet had nigh been lost on its way through the interior, owing to the treachery of an Indian. The fellow had undertaken to guide the Canadian servants of the Hudson Bay Company, who had it in charge, from York Factory to Cumberland House; but supposing, from its being unusual to forward packets at that season, that it must contain something of value, he seized an opportunity, when the two men had gone a little way from the river side, to steal the canoe, with its contents, and cross the river. There were no means of pursuit, and the poor men, destitute of food, without a gun, or even the means of making a fire, were obliged to march to the nearest establishment, through a very rugged and thickly-wooded country. They reached it after many days travelling, and much suffering, and as soon as they arrived, Mr. Mackintosh, the chief of the department, immediately sent off different parties in search of the culprit. They did not find him, though they got possession of the packet, which was torn open, and the letters scattered upon the ground, I need hardly mention that I afterwards remunerated the Ca-

4th.

Monday, 16th.

nadians for their sufferings and good conduct on this occasion.

Sunday, 22nd. On the morning of the 22nd we perceived a gray wolf crossing the lake, and Augustus and Ooligbuck went in pursuit. The speed of the animal, however, so much outstripped theirs, that it cooly halted to snap up an unfortunate fox which happened to cross its path, and bore it off in triumph. The visits of this animal were repeated for three successive days, and it at last became so bold as to steal fish, on two occasions, from a sledge which the dogs were accustomed to draw home from the nets, without a driver. The dogs were not touched, but this was accounted for when the wolf was killed, and found to be a female, as Mr. Dease informed us that at this season of the year the female wolves never attack the dog.

February. The month of February was a very anxious period of our winter's residence. The produce of the nets and fishing lines had been gradually diminishing during January, until the supply did not afford more than three or four of the small herrings per man; and none could be furnished to the dogs. The stock of dried meat was expended, and serious apprehensions were entertained of the party's suffering from want of food. The fish too, from being out of season, afforded very little nourishment, and frequent indisposition was the consequence with us all. Three of the stoutest men with whom this diet particularly disagreed, suffered very much from diarrhœa. It became, therefore, necessary to draw upon the stores of provision which had been set apart for the voyage along the sea-coast, and, on the 6th, we despatched three sledges to Fort Norman, for some pemmican, arrow root, and portable soup: they were likewise to bring any iron that could be procured from that establishment fit for being converted into nails or fastenings for the intended boat. This being the last opportunity of the season for forwarding letters to the southern department, I wrote to Governor Simpson and the council at York Factory, requesting that supplies of provisions might be stored for the Expedition, on the route to Canada and York Factory, and that the necessary means of conveyance might be provided for its return in 1827. All these arrangements requiring to be made a year in advance, I included the whole party in the estimate of the numbers to be provided for, that there might be no want of provision, if the western part of the Expedition should, from any cause, be obliged to retrace its steps. By the same conveyance I sent an account of our proceedings, with maps and drawings, to be forwarded to the Colonial Office.

On the 4th of this month, when all were heartily tired of short allowance, a report was brought of the traces of a moose deer having been seen about twelve miles from the fort. Had the days been longer, and a crust formed upon the snow, the hunters would have found no great difficulty in running down the animal, but our principal hope lay in their getting within shot without "raising it,"—the expression used when a deer is scared. Beaulieu being the most expert moose-hunter, went out on this occasion, accompanied by two others, Landré a Chipewyan lad, and a Dog-Rib hunter. When they arrived on the deer's track, they found that it had been raised, probably by the Indians who first discovered it; but anxious to procure meat for the fort, they commenced the pursuit. From their knowledge of the habits of the animal, and of the winding course it takes, they were enabled to shorten the distance; but after running four successive days without coming in sight, Beaulieu had the misfortune to fall over the stump of a tree, and sprain his ankle; the other two hunters being previously tired out. When this accident happened, they knew they were near the deer, and that it would soon give in, because its footsteps were stained with blood. Beaulieu, however, on account of his lameness, returned to the house, and his companions came with him. During the chase they bivouacked on the snow, and subsisted on a few ptarmigan which they killed. Landré after a night's rest, again set out, and was successful after two more day's running; not, however, without having nearly lost his life, for the moose, on receiving a shot, made a rush at him, striking furiously with his fore feet. He had just time to shelter himself behind a tree, upon which the animal spent its efforts, until his gun was again ready.

Saturday, 4th.

Landré's arrival with the joyful intelligence of his success, was hailed as the commencement of a season of plenty. When the moose meat was brought in, we had not an ounce of provision in store, and it was, therefore, most acceptable; although, from the manner in which it was hunted down, it proved exceedingly tough. In the evening, to increase our satisfaction, an Indian arrived with the information that the fish were plentiful at the station to which the Dog-Ribs had removed, and likewise that the hunters belonging to the fort had killed some rein-deer near their lodges. We immediately equipped four men with nets and lines, and sent them back with the Indian, giving them directions to report whether more persons could gain subsistence there. Their report, a few days afterwards, being favourable, four more

Friday, 10th.

men were despatched thither. They sent us some tittameg, weighing from six to eight pounds, which were the more acceptable, because none of that kind had been taken in our nets since the lake had been frozen over. By the time the moose was finished, the men came back from Fort Norman, with three bags of pemmican, which enabled us to continue the daily issue of rations, though the fare was still scanty.

Saturday, 25th. On the 25th, Beaulieu, accompanied by two men, went off in one direction, and the Dog-Rib hunters in another, in search of deer. Both parties were successful. From the former we received a summons, after four days' absence, to send sledges for meat, but not so from the Dog-Ribs, for they, to compensate for their long abstinence, consumed almost all the meat, and gorged themselves to such a degree, that they were unable to move, and became quite ill. From this period we had a sufficient supply of provision, because the fisheries improved, and we received deer from time to time. The men who had been indisposed gained strength, from the increased quantity, and amended quality, of their food; and we had also the gratification of seeing the dogs daily fatten, amidst the general plenty. The conduct of the men during the season of scarcity was beyond all praise; and the following anecdote is worthy of record, as displaying the excellent feeling of a British seaman, and as speaking the sentiments of the whole party. Talking with Robert Spinks as to the difference of his present food, from that to which he had been accustomed on board ship, I said I was glad the necessity was over of keeping them on short allowance. "Why, sir," said he, "we never minded about the short allowance, but were fearful of having to use the pemmican intended for next summer; we only care about the next voyage, and shall all be glad when the spring comes, that we may set off; besides, at the worst time, we could always spare a fish for each of our dogs." During the period of short allowance, the three dogs under the charge of this man were kept in better condition than any of the others.

We now called the men home from the nearest fishery, and set their nets near the Bear Lake River, but the men at the distant station with the Indians were kept there, and occasionally supplied the fort with fine tittameg and trout. The otters did considerable mischief to our nets at this time; six of these animals were seen in one day.

Many parheliæ were observed this month. On the 14th, at forty-five minutes after nine A.M., the arched form of the clouds, and the appearance of a collection of rays projected from the sun's disk in the shape of a fan, strongly resembled the coruscations of the aurora. The atmosphere was misty; temperature in the shade + 8° 5'; and when the thermometer with a blackened bulb was exposed to the sun's rays, it rose to + 43°. The magnetic needle, at nine A.M., was perceived to have made a greater deviation to the westward than usual at that hour, and I imagine that the cause of this increase probably arose from the atmosphere being then in a state of electricity, similar to that in which it is when the aurora appears in hazy weather; on which occasions we have observed that its coruscations have the strongest effect in causing aberrations of the needle. A violent gale from the north-west commenced on the 26th, and lasted, without intermission, for thirty-six hours.

March, 1st.

The early part of this month was marked by a succession of gales from the N.W., with a few intervals of moderate weather, in which the wind came from the east, and was attended by a clearer atmosphere than usually accompanies easterly winds in the colder months. We observed, with pleasure, on the 7th, that the sun had sufficient power to soften the snow in exposed places, and to form icicles from the roofs which had a southern aspect, but the return of strong winds from the W.N.W. brought back severe weather. On the 11th there was a violent gust of wind, which, in its passage over the lake, gathered up the snow in a column, similar to that of a waterspout.

Saturday, 11th.

Dr. Richardson made an excursion for the purpose of examining the rocks to the north of the establishment. He returned after two days, the snow being too deep for him to obtain specimens. The description he gave of a view from an eminence nine miles behind the fort, induced Lieutenant Back and me to visit the spot, and we were amply repaid for the walk. The view embraced the mountains on the borders of the Mackenzie to the west, a considerable portion of Bear Lake River, with the mountains near its rapids, Clark's Hill to the south, and the range of elevated land stretching to the east till they were lost in the distance. To the N.E. there appeared several small lakes, and the view was terminated by a portion of Bear Lake.

Tuesday, 21st.

The Chipewyan hunters who had been absent since Christmas, returned to us with their families, and brought with them a Dog-Rib girl, about twelve years old, who had been deserted by her tribe. When they found her, she was in the last stage of weakness, from famine, sitting by the expiring embers of a fire, and but for their timely appearance, death must soon have ended her sufferings. They fed and clothed her, and waited until she gained strength to accompany them. The wretches who had abandoned the poor creature, were on their way to a fishing station, which they knew to be very productive, and not above a day's march distant. She was unable to keep the pace at which they chose to proceed, and having no near relation but an aged aunt, who could not assist her, they left her at an encampment without any food. The hunters met this party of Indians about a month afterwards, when they were living in abundance. The girl, by that time, had perfectly recovered her strength, and they desired that she should be restored to them, but the hunters firmly resisted their importunity, and one of them adopted her as his own child. It is singular that she was the only female of the tribe that could be called good-looking. Her Indian name was Aton-larree, which the interpreter translated, Burnt-weed. Lieutenant Back made a sketch of her, in the dress which the hunter's wife gave to her on their first meeting. When the Indians came to the fort, I took the first opportunity of their being assembled in the hall, to send for the hunters and their wives, and to reward them by a substantial present of clothing and ammunition. I also gave to them some neat steel instruments, consisting of gimblets, and other useful articles, which they were desired to preserve, and show to other Indians, as a testimony of our approbation of their humanity. A present was also bestowed upon the girl, and then the Dog-Ribs were addressed as to their unfeeling conduct towards her. They listened quietly, and merely stated her weakness as the cause. There is little doubt but that the transactions of this day were canvassed afterwards, and it is to be hoped that the knowledge of our sentiments gaining circulation, may induce a discontinuance of their inhuman practices.

Wednesday, 22nd. By the men who had conveyed our last packet to Fort Simpson, we received intelligence that some Chipewyans had brought information to the Athabasca and Slave Lakes, of their having seen many indications of a party of white people on the sea coast eastward of the Coppermine River. The report stated, that they had found, in the preceding autumn, on the borders of a river near the sea-coast, a sawpit, some saws, and axes, and a store of deer's meat. There was snow on the ground, and the foot-steps of the party appeared recent. We concluded from these statements, that Captain Parry had laid up his ships in the vicinity of Bathurst's Inlet, and sent hunting-parties up the river to augment his stock of provision. I therefore despatched two men with letters to Mr. M'Vicar, at Slave Lake, containing a series of questions, that the matter might be thoroughly investigated, and requested him to transmit the answers to the Admiralty. I likewise begged of him immediately to procure a party of Indians to go to the spot, and convey a letter from me to Captain Parry, in order that they might either be employed as hunters for the ships, or carry their letters to the nearest establishment for conveyance to England. Had the information reached us sooner, so that a party could have gone from Bear Lake to the point at which the ships were, and returned before the men were wanted, I should have sent to ascertain the fact. The idea of the ships being on the northern coast, the prospect of their success, and the expectation of the eastern detatchment meeting them in the summer, afforded enlivening topics of conversation for several days, and on the day the intelligence came, we celebrated its arrival with a bowl of punch. The health of Captain Parry, and his party, as well as that of Captain Beechey, was drank with enthusiasm.

Thursday, 23rd.

We obtained observations for the time, from which it appeared that the chronometer, No. 1733, generously lent to the Expedition by my friend the late Mr. Moore, had only varied its rate two hundredths of a second, since the 3rd of November. I had worn it next my skin, suspended round my neck, the whole time; and, consequently, it was not exposed to much variation of temperature.

After the middle of this month the N.W. winds gave place to a succession of easterly breezes; whenever these prevailed, we observed the terrestrial refraction was much increased; double refraction of the land was not unfrequent, and twice the mist arising from the open water, appeared like a wall of ice. When the moon shone, halos, and occasionally paraselenæ, were visible; and towards the close of the month the coruscations of the aurora were often very brilliant.

During this month I noticed that on several occasions the magnetic needle oscillated when I approached it in a dress of water-proof cloth, although it remained stationary when others of the party examined it in their ordinary garments. The water-proof dress probably acted by exciting electricity in the body, although this opinion is rather contradicted by the fact of a fur cap, which had been rubbed by the hand until it affected the gold leaf electrometer, producing no change in the needle, and my approach to the electrometer not causing the gold-leaf to expand.

Having failed in an attempt to make charcoal for the blacksmith's use at this place, we despatched William Duncan, and the blacksmith, to make some at Fort Norman, where birch trees are plentiful; and on the 6th of April we were glad to see them return with the first load. The carpenters had already prepared the timbers and the keel for the new boat, and we were waiting for the coals to get the iron-work forward.

April, 6th.

Monday, 10th.

On the 10th Dr. Richardson and Mr. Kendall left the fort on snow-shoes, accompanied by an Indian guide, and a man driving a dog-sledge with provisions, for the purpose of completing the survey of Great Bear Lake, which Dr. Richardson had commenced the preceding autumn. The day was remarkably warm; the blackened thermometer, exposed to the sun, rose to + 90; and we hailed with delight a complete thaw. Cheered by the prospect, a spot was cleared of snow, the keel of the boat laid down, and that there might be no delay, all the sledges we could spare were despatched to fetch the remainder of the charcoal from Fort Norman. On the following day water was dripping from the roofs, and the flies were active within the rooms. The continuance of mild weather for six days caused a rapid decay of the snow, but no spots of land became visible. The men returned with the charcoal, and from them we learned that the season was more backward here than in the vicinity of Fort Norman. In the evening of the 17th, a telescope was put up in the meridian for finding the rates of the chronometers by the transit of Arcturus. On the 19th, thirty Hare-Indians arrived with sledges, bringing their winter's collection of furs for the Hudson Bay Company, and a large supply of dried meat for us, which, with the stock already in store, put us quite at ease respecting food until the season for our departure. The party consisted mostly of young lads, who, very good-naturedly, sang and danced for our amusement all the evening. They also gave us specimens of the dances in use among the Loucheux, which were more graceful than their own. The tune they sung to the Medicine-dance of the Loucheux, struck me as being soft and pretty. The ludicrous attitudes and grotesque figures of the dancers, as they wheeled in a circle, shaking the knives and feathers which they had between their fingers were happily sketched by Lieutenant Back.

Tuesday, 11th.

Wednesday, 19th.

As the fish had withdrawn from the open water at the commencement of the fine weather, the nets were brought nearer to the house; but we did not obtain more than thirty fish daily. This dimunition, however, gave us no concern, as we had plenty of meat. Shortly afterwards the trout began again to take bait, and we caught several of large size. Easterly winds prevailed this month, and they blew uninterruptedly from the 21st to the last day. A storm, on the 28th and 29th, delayed the carpenters working at the boat: the patches of ground which had for the last few days been visible, were again covered with snow, and the general aspect was bleak and wintry.

Dr. Richardson and Mr. Kendall returned on the 1st of May, and we were furnished with the following particulars of their journey. Their course, on leaving us, was first directed to the fishery in Mac Vicar's Bay, which they reached on the fourth day, and from whence, taking with them another sledge-load of provisions and an additional attendant, they continued their journey to the bottom of Mac Tavish Bay, the most easterly part of the Lake. The reduction in their stock of provisions now caused them to commence their return, and they reached the fort after an absence of three weeks, during which, in very unfavourable weather, they travelled about three hundred and eighty miles. Dr. Richardson had sailed four hundred and eighty miles through the lake in the autumn, and in the two excursions, five hundred miles of its shores were delineated, and the positions of many points established by astronomical observations. About twenty miles of the north shore of Mac Tavish Bay are the only parts of the Bear Lake remaining unsurveyed. The following brief description of Bear Lake is extracted from Dr. Richardson's Journal:—

May, 1st.

"Great Bear Lake is formed by the union of five arms or bays, which were named after Messrs. Keith, Smith, Dease, Mac Tavish, and Mac Vicar, of the Hudson's Bay Company. The principal feeding-stream, named Dease River, rises in the Copper Mountains, and falls into the upper end of Dease Bay, which is the most northern part of the lake, and Bear Lake River, which conveys the waters of the lake to the Mackenzie, issues from Keith Bay, the most southerly arm. Mac Tavish Bay is the most easterly portion of the lake, and Smith Bay, which lies opposite to it, runs to the westward. Mac Vicar Bay has a southerly direction nearly parallel to Keith Bay. The length of the lake, from Dease River to Bear Lake River, is about one hundred and seventy-five miles; and its breadth, from the bottom of Smith Bay to the bottom of Mac Tavish Bay, is one hundred and fifty miles. A range of granite hills skirts the bottom of Mac Tavish Bay. The Great Bear Mountain, at whose base some bituminous shale cliffs are exposed, is about nine hundred feet high, and separates Mac Vicar and Keith Bays; a similar mountain lies betwixt Keith and Smith Bays. In Dease Bay, limestone and sandstone are the prevailing rocks. The waters of the lake are very clear, and of unknown depth; forty-five fathoms of line were let down near the shore, in Mac Tavish Bay, without reaching the bottom. There is a considerable quantity of good wood, principally white spruce, in the vicinity of the lake; but there is reason to believe that, before many years elapse, it will become scarce, for it is very slow of growth, and the natives every year set fire to it in various quarters, and thus destroy it for many miles. The finest timber was observed on the west side of Great Bear Lake Mountain. There are good fisheries in Dease Bay, and in various other quarters of the lake; but the fish taken in Mac Vicar Bay are remarkably fine and abundant at all seasons of the year. The principal advantage of the site chosen for Fort Franklin, is its vicinity to the Bear Lake River, and the great quantity of fish that can be procured at certain seasons, although they are small and of inferior quality."

On the 5th of this month, the men being called in from the fishery in Mac Vicar Bay, the whole party was once more assembled at the house, anxiously looking forward to the arrival of spring. We hailed the appearance of swans, on the following day, as a sure sign of its approach. A goose was seen on the 7th, two ducks on the 8th, and on the 9th several gulls were observed in the open water near the Bear Lake River. The snow, at this time, was rapidly diminishing from the surface of the lake, and there were many spots of ground visible. We, therefore, commenced the preparations for the summer's voyage. The seamen were employed in repairing the coverings and sails of the boats, as well as in refitting their rigging, and occupation was allotted to every person in the establishment. These operations requiring the constant superintendence of the officers, the observations of the magnetic-needle were discontinued. After the middle of the month, we were visited by occasional showers of rain, which removed the snow, and produced a perceptible decay of the ice.

On the 23d, the ice broke away from the shore of the small lake, and also of Bear Lake, in front of the house. Swans and geese were now daily passing to the northward; many shots were fired at them, both by the Indians and our own party, but only a few were killed. The geese were principally of the kind known to naturalists by the name of Canada geese, and denominated bustards by the voyagers. Numbers of white geese also passed; we saw only two flocks of laughing-geese. The first swallow came on the 16th, and, on the following day, many others arrived. A variety of ducks, gulls, and many of the small aquatic birds, now frequented the marshy borders of the little lake, which afforded constant amusement to the sportsmen, and full occupation to Dr. Richardson in preparing the skins for specimens.

Wednesday, 24th.

On the 24th, the musquitoes appeared, feeble at first, but, after a few days, they became vigorous and tormenting. The first flower, a tussilago, was gathered on the 27th. Before the close of the month, several others were in bloom, of which the most abundant was the white anemone (*anemone tenella.*) The leaf-buds had not yet burst, though just ready to open.

The carpenters had now finished the new boat, which received the name of the Reliance. It was constructed of fir, with birch timbers, after the model of our largest boat, the Lion, but with a more full bow, and a finer run abaft. Its length was about twenty-six feet, and breadth five feet eight inches. It was fastened in the same manner as the other boats, but with iron instead of copper, and to procure sufficient nails we were obliged to cut up all the spare axes, trenches*, and ice-chisels. Being without tar, we substituted strips of water-proof canvass, soaked in some caoutchouc varnish, which we had brought out, to lay between the seams of the planks; and for paint, we made use of resin, procured from the pine-trees, boiled and mixed with grease. The other boats were afterwards put in complete repair. The Lion required the most, in consequence of the accident in Bear Lake River. The defects in the other two principally arose from their having been repaired at Cumberland House with the elm that grows in its vicinity, and is very spongy. We now substituted white spruce fir, which, when grown in these high latitudes, is an excellent wood for boat-building. We were surprised to find, that, notwithstanding the many heavy blows these boats had received in their passage to this place, there was not a timber that required to be changed.

In our bustle, we would gladly have dispensed with the presence of the Dog-Ribs, who now visited us in great numbers, without bringing any supplies. They continued hanging about the fort, and their daily drumming and singing over the sick, the squalling of the children, and bawling of the men and women, proved no small annoyance. We were pleased, however, at perceiving that the ammunition we had given to them in return for meat, had enabled them to provide themselves with

* Used by the Indians to break up the beaver lodges.

leathern tents. Their only shelter from the wind, snow, or rain, before this season, had been a rude barricade of pine branches. Fortunately, for our comfort, they were obliged to remove before the expiration of the month to a distant fishery to procure provision.

June, 1st. The preparations for the voyage along the coast being now in a state of forwardness, my attention was directed to the providing for the return of Dr. Richardson's party to this establishment in the following autumn, and to the securing means of support for all the members of the Expedition at this place, in the event of the western party being likewise compelled to return to it. Respecting the first point, it was arranged that Beaulieu the interpreter, and four Canadians, should quit Fort Franklin on the 6th of August, and proceed direct to Dease River with a bateau, and wait there until the 20th of September, when, if Dr. Richardson did not appear, they were to come back to the fort in canoes, and to leave the boat, with provision and other necessaries, for the use of the eastern detachment. All these points were explained to Beaulieu, and he not only understood every part of the arrangement, but seemed very desirous to perform the important duty entrusted to him. I next drew up written instructions for the guidance of Mr. Dease, during the absence of the Expedition, directing his attention first to the equipment and despatch of Beaulieu on the 6th of August, and then to the keeping the establishment well stored with provision. He was aware of the probability that the western party would meet his Majesty's ship Blossom, and go to Canton in her. But as unforeseen circumstances might compel us to winter on the coast, I considered it necessary to warn him against inferring, from our not returning in the following autumn, that we had reached the Blossom. He was, therefore, directed to keep Fort Franklin complete, as to provision, until the spring of 1828. Dr. Richardson was likewise instructed, before he left the fort in 1827, on his return to England, to see that Mr. Dease fully understood my motives for giving these orders, and that he was provided with the means of purchasing the necessary provision from the Indians.

Wednesday, 7th. The long reign of the east wind was at length terminated by a fresh N.W. breeze, and the ice yet remaining on the small lake soon disappeared, under the softening effects of this wind. This lake had been frozen eight months, wanting three days. A narrow channel being opened along the western border of Bear Lake, on the 14th Dr. Richardson took advantage of it, and went in a small canoe with two men to examine the mountains on the borders of Bear Lake River, and to collect specimens of the plants that were now in flower, intending to rejoin the party at Fort Norman. On the same day, in 1821, the former Expedition left Fort Enterprize for the sea.

Thursday, 15th.

The equipments of the boats being now complete, they were launched on the small lake, and tried under oars and sails. In the afternoon the men were appointed to their respective stations, and furnished with the sky-blue waterproof uniforms, and feathers, as well as with the warm clothing which had been provided for the voyage. I acquainted them fully with the object of the Expedition, and pointed out their various duties. They received these communications with satisfaction, were delighted with the prospect of the voyage, and expressed their readiness to commence it immediately. Fourteen men, including Augustus, were appointed to accompany myself and Lieutenant Back, in the Lion and Reliance, the two larger boats; and ten, including Ooligbuck, to go with Dr. Richardson and Mr. Kendall, in the Dolphin and Union. In order to make up the complement of fourteen for the western party, I proposed to receive two volunteers from the Canadian voyagers; and to the credit of Canadian enterprise, every man came forward. I chose François Felix and Alexis Vivier, because they were the first who offered their services, and this too without any stipulation as to increase of wages.

Spare blankets, and every thing that could be useful for the voyage, or as presents to the Esquimaux, which our stores could furnish, were divided between the eastern and western parties, and put up into bales of a size convenient for stowage. This interesting day was closed by the consumption of a small quantity of rum, reserved for the occasion, followed by a merry dance, in which all joined with great glee, in their working dresses. On the following Sunday the officers and men assembled at Divine service, dressed in their new uniforms; and in addition to the ordinary service of the day, the special protection of Providence was implored on the enterprise we were about to commence. The guns were cleaned the next day, and stowed in the arm chests, which had been made to fit the boats. Tuesday and Wednesday were set apart for the officers and men to pack their own things. A strong western breeze occurred on the 21st, which removed the ice from the front of the house and opened a passage to the Bear Lake River. The men were sent with the boats and stores to the river in the evening, and were heartily cheered on quitting the beach.

The officers remained to pack up the charts, drawings, and other documents, which were to be left at the fort; and, in the event of none of the officers returning, Mr. Dease was directed to forward them to England. We quitted the house at half past ten, on Thursday morning, leaving Coté, the fisherman, in charge, until Mr. Dease should return from Fort Norman. This worthy old man, sharing the enthusiasm that animated the whole party, would not allow us to depart without giving his hearty, though solitary cheer, which we returned in full chorus.

The position of Fort Franklin was determined to be in latitude 65° 11′ 56″ N., longitude 123° 12′ 44″ W.; variation of the compass 39° 9′ E.; dip of the needle 82° 58′ 15″.

CHAPTER III.

Voyage to the Sea—Part from the Eastern Detachment at Point Separation—Reach the Mouth of the Mackenzie—Interview and Contest with the Esquimaux—Detained by Ice—Meet friendly Esquimaux—Point Sabine.

On our arrival at the Bear Lake River, we were mortified to find the ice drifting down in large masses, with such rapidity as to render embarkation unsafe. The same cause detained us the whole of the following day; and as we had brought no more provision from the house than sufficient for an uninterrupted passage to Fort Norman, we sent for a supply of fish. This was a very sultry day, the thermometer in the shade being 71° at noon, and 74° at three P.M.

Thursday, 22nd.

The descent of the ice having ceased at eight in the morning of the 24th, we embarked. The heavy stores were put into a bateau, manned by Canadians, who were experienced in the passage through rapids, and the rest of the boats were ordered to follow in its wake, keeping at such a distance from each other as to allow of any evolution that might be necessary to avoid the stones. The boats struck several times, but received no injury. At the foot of the rapid we met a canoe, manned by four of our Canadian voyagers, whom Dr. Richardson had sent with some letters that had arrived at Fort Norman from the Athabasca Lake; and as the services of the men were wanted, they were embarked in the boats, and the canoe was left. Shortly afterwards we overtook Beaulieu, who had just killed a young moose deer, which afforded the party two substantial meals. At this spot, and generally along the river, we found abundance of wild onions.

We entered the Mackenzie River at eight in the evening, and the current being too strong for us to advance against the stream with oars, we had recourse to the tracking line, and travelled all night. It was fatiguing, owing to large portions of the banks having been overthrown by the disruption of the ice, and from the ground being so soft that the men dragging the rope sank up to the knees at every step; but these impediments were less regarded than the ceaseless torment of the musquitoes. We halted to sup at the spot where Sir A. Mackenzie saw the flame rising from the bank in 1789. The precipice was still on fire, the smoke issuing through several apertures. Specimens of the coal were procured.

Sunday, 25th. We reached Fort Norman at noon on the 25th. On the following morning the provision and stores which had been left at this place were examined, and found to be in excellent order, except the powder in one of the magazines, which had become caked from damp. I had ordered a supply of iron-work, knives, aud beads, for the sea voyage from Fort Simpson; they had arrived some days before us, and with our stock thus augmented, we were well furnished with presents for the natives. The packages being finished on the 27th, the boats received their respective ladings, and we were rejoiced to find that each stowed her cargo well, and with her crew embarked floated as buoyantly as our most sanguine wishes had anticipated. The heavy stores, however, were afterwards removed into a bateau that was to be taken to the mouth of the river, to prevent the smaller boats from receiving injury in passing over the shoals.

We waited one day to make some pounded meat we had brought into pemmican. In the mean time the seamen enlarged the foresail of the Reliance.

The letters which I received from the Athabasca department informed me that the things I had required from the Company in February last, would be duly forwarded; they likewise contained a very different version of the story which had led us to suppose that Captain Parry was passing the winter on the northern coast. We now learned that the Indians had only seen some pieces of wood recently cut, and a deer that had been killed by an arrow; these things we concluded were done by the Esquimaux. Three men from Slave Lake, whom I had sent for to supply the place of our Chipewyan hunters, who were very inactive last winter, joined us at this place. They were to accompany Mr. Dease and the Canadians to Fort Franklin; and that they, as well as the Indians, might have every encouragement to exert themselves in procuring provisions during the summer, I directed a supply of the goods they were likely to require, to be sent from Fort Simpson, as soon as possible. The longitude of Fort Norman was observed to be 124° 44′ 47″ W., its latitude 64° 40′ 38″ N.; variation 39° 57′ 52″ E.

Wednesday, 28th. Early this morning the boats were laden and decorated with their ensigns and pendants, and after breakfast we quitted the fort, amidst the hearty cheers of our friends Mr. Dease, Mr. Brisbois, and the Canadians, and I am sure carried their best wishes for our success.

We halted at noon to obtain the latitude, which placed the entrance of Bear Lake River in 64° 55′ 37″ N.; and Dr. Richardson took advantage of this delay to visit the mountain at that point, but his stay was short, in consequence of a favourable breeze springing up. We perceived that the four boats sailed at nearly an equal rate in light breezes, but that in strong winds the two larger ones had the advantage. When we landed to sup the musquitoes beset us so furiously that we hastily despatched the meal and re-embarked, to drive under easy sail before the current. They continued, however, to pursue us, and deprived us of all rest. On our arrival, next morning, at the place of the first rapids, there was scarcely any appearance of broken water, and the sand-bank on which Augustus had been so perilously situated in the preceding autumn, was entirely covered. This was, of course, to be ascribed to the spring floods; the increase of water to produce such a change, must have exceeded six feet. In the afternoon we were overtaken by a violent thunder-storm, with heavy rain, which made us apprehensive for the pemmican, that spoils on being wet. It unfortunately happened that a convenient place for spreading out the bags that were injured could not be found, until we reached the Hare-Skin River, below the Rampart Defile, which was at nine o'clock. They were spread out the next morning, with the other perishable parts of the cargo, and we remained until they were dry. We embarked at ten, and, aided by a favourable breeze, made good progress until six P.M., when the threatening appearance of the clouds induced us to put on shore, and we had but just covered the baggage before heavy rain fell, that continued throughout the night. Four Hare Indians came to the encampment, to whom dried meat and ammunition were given, as they were in want of food from being unable to set their nets in the present high state of the water. These were the only natives seen since our departure from Fort Norman; they informed us, that, in consequence of not being able to procure a sufficiency of fish in the Mackenzie at this season, their companions had withdrawn to gain their subsistence from the small lakes in the interior.

Friday, 30th.

We embarked at half past one on the morning of the 1st of July. The sultry weather of the preceding day made us now feel more keenly the chill of a strong western breeze, and the mist which it brought on, about four hours after our departure. This wind being contrary to the current, soon raised such high waves that the boats took in a great deal of water; and as we made but little progress, and

July, 1st.

were very cold, we landed to kindle a fire, and prepare breakfast; after which we continued the voyage to Fort Good Hope, without any of the interruptions from sand-banks that we had experienced in the autumn.

On our arrival we were saluted with a discharge of musketry by a large party of Loucheux, who had been some time waiting at the fort, with their wives and families, for the purpose of seeing us. After a short conference with Mr. Bell, the master of the post, we were informed that these Indians had lately met a numerous party of Esquimaux at the Red River, by appointment, to purchase their furs; and that in consequence of a misunderstanding respecting some bargain, a quarrel had ensued between them, which fortunately terminated without bloodshed. We could not, however, gain any satisfactory account of the movements of the Esquimaux. The only answers to our repeated questions on these points were, that the Esquimaux came in sixty canoes to Red River, and that they supposed them to have gone down the eastern channel, for the purpose of fishing near its mouth. The chief, however, informed us that he had mentioned our coming to their lands this spring, and that they had received the intelligence without comment; but from his not having alluded to this communication until the question was pressed upon him, and from the manner of his answering our inquiries, I thought it doubtful whether such a communication had really been made.

We had been led to expect much information from the Loucheux respecting the channels of the river, and the coast on the east and west side near its mouth, but we were greatly disappointed. They were ignorant of the channel we ought to follow in order to arrive at the western mouth of the river; and the only intelligence they gave us respecting the coast on that side was, that the Esquimaux represented it to be almost constantly beset by ice. They said also that they were unacquainted with the tribes who reside to the westward. Several of the party had been down the eastern channel, of which they made a rude sketch; and their account of the coast on that side was, that, as far as they were acquainted with it, it was free from ice during the summer.

Mr. C. Dease, the former master of Fort Good Hope, had retained two of the Loucheux to accompany the Expedition until we should meet the Esquimaux: they spoke a few words of the language, which they had learned during an occasional residence with the tribe that resides on the eastern border of the river. But the knowledge of the recent transactions at Red River had convinced us that their presence would be more likely to irritate than pacify the Esquimaux. We also discovered that their sole motive for accompanying us was the desire of trading with that people; and further, that they expected we should take their families and baggage in the boats. Their services were therefore declined; and a compensation was offered to them for their loss of time in waiting for us; but having fixed their minds on the gain to be derived from us and from the trade with the Esquimaux, they expressed great disappointment, and were very intemperate in their language. As I was anxious, for the sake of the trade at the post, to leave them no room to complain either of us, or of Mr. Dease who had acted for us, I spent several hours in debate with them to very little purpose, and at last discovered that the whole scene was got up for the purpose of obtaining a few more goods. My compliance with their wish rendered them quite contented. I afterwards added a present to the principal chief of the party, who still expressed a wish to accompany us, but he frankly said that if he went, all his young men must go also. They came in the evening in great good humour to exhibit their dances in front of our tent, a compliment we could well have dispensed with, as we were busy.

Having ascertained that the Esquimaux were likely to be seen in greater numbers than had been at first imagined, I increased the stock of presents from the store at this place, and exchanged two of our guns, which were defective, that the party might have entire confidence in their arms. And to provide against the casualty of either or both branches of the Expedition having to return this way, I requested Mr. Bell to store up as much meat as he could during the summer. We learned from this gentleman that the supply of meat at this post was very precarious, and that had we not left the five bags of pemmican in the autumn, the residents would have been reduced to great distress for food during the winter. These bags were now replaced. The arrangements being concluded, we spent the greater part of the night in writing to England. I addressed to the Colonial Secretary an account of our proceedings up to this time, and I felt happy to be able to state that we were equipped with every requisite for the Expedition.

Sunday, 2nd.

We quitted Fort Good Hope at five on the 2nd. In the passage down the river we were visited by several Loucheux, who, the instant we appeared, launched their canoes, and came off to welcome us. We landed, at their request, to purchase fish; yet, after the bargain had been completed, an old woman stepped forward, and would only allow of our receiving two fish: she maintained her point, and carried off the rest in spite of all remonstrance. The natives were all clothed in new leathern dresses, and looked much neater, and in better health, than last autumn. Being anxious to reach the Red River, we continued rowing against the wind until after midnight. On reaching that place, the ground proved too wet for us to encamp; we, therefore, proceeded a short distance lower down, and put up under some sandstone cliffs, where there was but just room for the tents. As we were now on the borders of the Esquimaux territory, we devoted the following morning to cleaning the arms; and a gun, dagger, and ammunition, were issued to each person. We had no reason, indeed, to apprehend hostility from the Esquimaux, after the messages they had sent to Fort Franklin, but vigilance and precaution are never to be omitted in intercourse with strange tribes.

Monday, 3rd. Embarking at two in the afternoon of the 3rd, we soon entered the expansion of the river whence the different channels branch off, and steering along the western shore, we came to the head of a branch that flowed towards the Rocky Mountain range. Being anxious not to take the eastern detachment out of their course, I immediately encamped to make the necessary arrangements for the separation of the parties. The warm clothing, shoes, and articles for presents, had been previously put up in separate packages, but the provisions remained to be divided, which was done in due proportion. Twenty-six bags of pemmican, and two of grease, were set apart for the Dolphin and Union, with a supply of arrow-root, macaroni, flour, and portable soup, making in all eighty days' provision, with an allowance for waste. The Lion and Reliance received thirty-two bags of pemmican, and two of grease, with sufficient arrow-root, &c., to make their supply proportionate to that of the eastern party. Provided no accident occurred, neither party could be in absolute want for the whole summer, because at two-thirds allowance the pemmican could be made to last one hundred days; and we had reason to expect to meet with deer occasionally.

In the evening I delivered my instructions to Dr. Richardson; they were in substance as follows:—He was to take under his charge Mr. E. N. Kendall, and ten men, and proceed in the Dolphin and Union to survey the coast between the Mackenzie and Copper-Mine Rivers. On reaching the latter river, he was to travel by land to the northeast arm of Great Bear Lake, where Beaulieu was under orders to meet him with a boat for the conveyance of his party to Fort Franklin. But if he should be so much delayed on the coast as to have no prospect of reaching the Copper-Mine River by the close of August, or the Bear Lake Portage by the 20th of September, he was not to expose himself or his party to risk by persevering beyond the 15th or 20th of August, but was to return to Fort Franklin by way of the Mackenzie, or by any other route he might discover. The only cause of regret I had respecting the equipment of the eastern party was my being unable to provide Dr. Richardson with a chronometer, the main-springs of two out of the three chronometers furnished to us having been broken. I borrowed, however, from Mr. Dease, a watch, made by Barraud, to enable Mr. Kendall to obtain the longitude by lunar distances. They were likewise provided with that excellent instrument Massey's Log; and knowing Mr. Kendall's intimate acquaintance with marine surveying, I had no doubt of his being able to make a correct survey of the coast. The spot where the above arrangements were made, bears the name of Point Separation, and lies in latitude 67° 38′ N., longitude 133° 53′ W.

As the parties entertained for each other sentiments of true friendship and regard, it will easily be imagined that the evening preceding our separation was spent in the most cordial and cheerful manner. We felt that we were only separating to be employed on services of equal interest; and we looked forward with delight to our next meeting, when, after a successful termination, we might recount the incidents of our respective voyages. The best supper our means afforded was provided, and a bowl of punch crowned the parting feast.

We were joined by an elderly Loucheux, who gave us a better account of the eastern and western channels than we had hitherto obtained. "The west branch," he said, "would take us to the sea, and flowed the whole way at no great distance from the mountains." "The eastern was a good channel, and passed close to the hills on that side." He further informed us that the Esquimaux were generally to be found on an island in the eastern channel, but were seldom seen in the western branch. He was, however, unacquainted with the coast, and we found afterwards that he knew little about the movements of the Esquimaux.

Tuesday, 4th.

By six in the morning of the 4th the boats were all laden, and ready for departure. It was impossible not to be struck with the difference between our present complete state of equipment and that on which we had embarked on our former disastrous voyage. Instead of a frail bark canoe, and a scanty supply of food, we were now about to commence the sea voyage in excellent boats, stored with three months' provision. At Dr. Richardson's desire the western party embarked first. He and his companions saluted us with three hearty cheers, which were warmly returned; and as we were passing round the point that was to hide them from our view, we perceived them also embarking. Augustus was rather melancholy, as might have been expected, on his parting from Ooligbuck, to proceed he knew not whither; but he recovered his wonted flow of spirits by the evening.

The western party were distributed as follows:—

LION.	RELIANCE.
John Franklin, *Captain R. N.*	George Back, *Lieutenant R. N.*
William Duncan, *Cockswain.*	Robert Spinks, *Cockswain.*
Thomas Matthews, *Carpenter.*	Robert Hallom, *Corpl. of Marines.*
Gustavus Aird, *Bowman.*	Charles Mackenzie, *Bowman.*
George Wilson, *Marine.*	Alexander Currie, *Middle Man.*
Archibald Stewart, *Soldier.*	Robert Spence, *Ditto.*
Neil Mac Donald, *Voyager.*	Alexis Vivier, *Canadian.*
Augustus, *Esquimaux.*	Francois Felix, *Ditto.*

Our course was directly towards the Rocky Mountain range, till we came near the low land that skirts its base; where, following the deepest channel, we turned to the northward. I was desirous of coasting the main shore, but finding some of the westermost branches too shallow, we kept on the outside of three islands for about twelve miles, when we entered the channel that washes the west side of Simpson's Island. It was winding, and its breadth seldom exceeded a quarter of a mile. During our progress we occasionally caught a glimpse of the Rocky Mountains, which was an agreeable relief to the very dull picture that the muddy islands in our neighbourhood afforded. We halted to breakfast just before noon, and observed the latitude 67° 51′ N.

In the afternoon one deer was seen, and many swans and geese; we did not fire at them, for fear of alarming any Esquimaux that might be near. Encamped at eight P.M., opposite Simpson's Island, in latitude 68° 13′ N., longitude 134° 27′ W. The boats were secured without discharging the cargoes, and two men were placed on guard, to be relieved every two hours.

Wednesday, 5th. We set forward at four A.M., with a favourable breeze, and made good progress, though the river was very winding. At eight we entered a branch that turned to the westward round the point of Halkett Island into the channel washing the main shore. We soon afterwards arrived at a spot where a large body of Esquimaux had been encamped in the spring, and supposing that they might revisit this place, a present of an ice-chisel, kettle, and knife, was hung up in a conspicuous situation. Soon after we had entered the channel that flows by the main shore, we first perceived lop-sticks, or pine trees, divested of their lower branches, for the purpose of land-marks, and therefore concluded it was much frequented by the Esquimaux. Our course was then altered to N.W., and we soon passed the last of the well-wooded islands. The spruce fir-trees terminated in latitude 68° 36′ N; and dwarf willows only grew below this part. A very picturesque view was obtained of the Rocky Mountains, and we saw the entire outline of their peaked hills, table-land, and quoin shaped terminations. Two lofty ranges were fronted by a lower line of round-backed hills, in which we perceived the strata to be horizontal, and the stone of a yellow colour. A few miles lower down we found hills of sand close to the west border of the river. We passed several deserted huts, and in one spot saw many chips and pieces of split drift-wood, that appeared to have been recently cut. The channel varied in breadth from a half to three-quarters of a mile, but, except in the stream of the current, the water was so shallow as scarcely to float the boats, and its greatest depth did not exceed five feet. We landed at eight P.M., on Halkett Island, intending to encamp, but owing to the swampiness of the ground the tent could not be pitched. Having made a fire and cooked our supper, we retired to sleep under the coverings of the boats, which afforded us good shelter from a gale and heavy rain that came on before midnight. Latitude 68° 39′ N., longitude 135° 35′ W.

Thursday, 6th.

The continuance of stormy weather detained us until two P.M. of the 9th, when the rain ceasing, we embarked. After passing through the expansion of the river near the west extreme of Halkett Island, we turned into the narrower and more winding channel, between Colvill Island and the main. A fog coming on at eight P.M. we encamped, in latitude 68° 48′ N., longitude 136° 4′ W.; temperature of the air 42°, that of the water being 47° in the middle of the stream. Several of the glaucous gulls were seen, and this circumstance, as well as a line of bright cloud to the N.W. resembling the ice-blink, convinced us that the sea was not far off. A rein-deer appearing near the encampment, two men were sent after it, who returned unsuccessful. Augustus obtained a goose for supper. Many geese, swans, and ducks, had been seen on the marshy shores of the island in the course of the day.

Friday, 7th. The night was cold, and at day-light on the 7th the thermometer indicated 36°. Embarking at four A.M. we sailed down the river for two hours, when our progress was arrested by the shallowness of the water. Having endeavoured, without effect, to drag the boats over the flat, we remounted the stream to examine an opening to the westward, which we had passed. On reaching the opening we found the current setting through it into the Mackenzie, by which we knew that it could not afford a passage to the sea, but we pulled up it a little way, in the hope of obtaining a view over the surrounding low grounds from the top of an Esquimaux house which we saw before us. A low fog, which had prevailed all the morning, cleared away, and we discovered that the stream we had now ascended issued from a chain of lakes lying betwixt us and the western hills, which were about six miles distant, the whole intervening country between the hills and the Mackenzie being flat.

After obtaining an observation for longitude in 136° 19′ W., and taking the bearing of several remarkable points of the Rocky Mountain range, we returned to the Mackenzie, and passing the shallows which had before impeded us, by taking one half the boats' cargoes over at a time, we came in sight of the mouth of the river. Whilst the crews were stowing the boats, I obtained an observation for latitude in 68° 53′ N., and having walked towards the mouth of the river, discovered on an island, which formed the east side of the bay into which the river opened, a crowd of tents, with many Esquimaux strolling amongst them. I instantly hastened to the boats, to make preparations for opening a communication with them, agreeably to my instructions. A selection of articles for presents and trade being made, the rest of the lading was closely covered up; the arms were inspected, and every man was directed to keep his gun ready for immediate use. I had previously informed Lieutenant Back of my intention of opening the communication with the Esquimaux by landing amongst them, accompanied only by Augustus; and I now instructed him to keep the boats afloat, and the crews with their arms ready to support us in the event of the natives proving hostile; but on no account to fire until he was convinced that our safety could be secured in no other way. Having received an impression from the narratives of different navigators that the sacrifices of life which had occurred in their interviews with savages, had been generally occasioned by the crews mistaking noise and violent gestures for decided hostility, I thought it necessary to explain my sentiments on this point to all the men, and peremptorily forbade their firing till I set the example, or till they were ordered to do so by Lieutenant Back. They were also forbidden to trade with the natives on any pretence, and were ordered to leave every thing of that kind to the officers.

On quitting the channel of the river we entered into the bay, which was about six miles wide, with an unbounded prospect to seaward, and steered towards the tents under easy sail, with the ensigns flying. The water became shallow as we drew towards the island, and the boats touched the ground when about a mile from the beach; we shouted, and made signs to the Esquimaux to come off, and then pulled a short way back to await their arrival in deeper water. Three canoes instantly put off from the shore, and before they could reach us others were launched in such quick succession, that the whole space between the island and the boats was covered by them. The Esquimaux canoes contain only one person, and are named *kaiyacks;* but they have a kind of open boat capable of holding six or eight people, which is named *oomiak.* The men alone use the kaiyacks, and the oomiaks are allotted to the women and children. We endeavoured to count their numbers as they approached, and had proceeded as far as seventy-three canoes, and five oomiaks, when the sea became so crowded by fresh arrivals, that we could advance no farther in our reckoning. The three headmost canoes were paddled by elderly men, who, most probably, had been selected to open the communication. They advanced towards us with much caution, halting when just within speaking distance, until they had been assured of our friendship, and repeatedly invited by Augustus to approach and receive the present which I offered to them. Augustus next explained to them in detail the purport of our visit, and told them that if we succeeded in finding a navigable channel for large ships, a trade highly beneficial to them would be opened. They were delighted with this intelligence, and repeated it to their countrymen, who testified their joy by tossing their hands aloft, and raising the most deafening shout of applause I ever heard.

After the first present, I resolved to bestow no more gratuitously, but always to exact something, however small, in return; the three elderly men readily offered the ornaments they wore in their cheeks, their arms, and knives, in exchange for the articles I gave them. Up to this time the first three were the only kaiyacks that had ventured near the boats, but the natives around us had now increased to two hundred and fifty or three hundred persons, and they all became anxious to share in the lucrative trade which they saw established, and pressed eagerly upon us, offering for sale their bows, arrows, and spears, which they had hitherto kept concealed within their canoes. I endeavoured in vain, amidst the clamour and bustle of trade, to obtain some information respecting the coast, but finding the natives becoming more and more importunate and troublesome, I determined to leave them, and, therefore, directed the boats' heads to be put to seaward. Notwithstanding the forwardness of the Esquimaux, which we attributed solely to the desire of a rude people to obtain the novel articles they saw in our possession, they had hitherto shown no unfriendly disposition; and when we told them of our intention of going to sea, they expressed no desire to detain us, but, on the contrary, when the Lion grounded in the act of turning, they assisted us in the kindest manner by dragging her round. This manœuvre was not of much advantage to us, for, from the rapid ebbing of the tide, both boats lay aground; and the Esquimaux told us, through the medium of Augustus, that the whole bay was alike flat, which we afterwards found to be correct.

An accident happened at this time, which was productive of unforeseen and very annoying consequences. A kaiyack being overset by one of the Lion's oars, its owner was plunged into the water with his head in the mud, and apparently in danger of being drowned. We instantly extricated him from his unpleasant situation, and took him into the boat until the water could be thrown out of his kaiyack, and Augustus, seeing him shivering with cold, wrapped him up in his own great coat. At first he was exceedingly angry, but soon became reconciled to his situation, and looking about, discovered that we had many bales, and other articles, in the boat, which had been concealed from the people in the kaiyacks, by the coverings being carefully spread over all. He soon began to ask for every thing he saw, and expressed much displeasure on our refusing to comply with his demands; he also, as we afterwards learned, excited the cupidity of others by his account of the inexhaustible riches in the Lion, and several of the younger men endeavoured to get into both our boats, but we resisted all their attempts. Though we had not hitherto observed any of them stealing, yet they showed so much desire to obtain my flag, that I had it furled and put out of sight, as well as every thing else that I thought could prove a temptation to them. They continued, however, to press upon us so closely, and made so many efforts to get into the boats, that I accepted the offer of two chiefs, who said that if they were allowed to come in, they would keep the others out. For a time they kept their word, and the crews took advantage of the respite thus afforded, to endeavour to force the boats towards the river into deeper water. The Reliance floated, but the Lion was immoveable, and Lieutenant Back dropping astern again made his boat fast to the Lion by a rope. At this time one of the Lion's crew perceived that the man whose kaiyack had been upset had a pistol under his shirt, and was about to take it from him, but I ordered him to desist, as I thought it might have been purchased from the Loucheux. It had been, in fact, stolen from Lieutenant Back, and the thief, perceiving our attention directed to it, leaped out of the boat, and joined his countrymen, carrying with him the great coat which Augustus had lent him.

The water had now ebbed so far, that it was not knee deep at the boats, and the younger men wading in crowds around us, tried to steal every thing within their reach; slyly, however, and with so much dexterity, as almost to escape detection. The moment this disposition was manifested, I directed the crews not to suffer any one to come alongside, and desired Augustus to tell the two chiefs, who still remained seated in the Lion, that the noise and confusion occasioned by the crowd around the boats greatly impeded our exertions; and that if they would go on shore and leave us for the present, we would hereafter return from the ship which we expected to meet near this part of the coast, with a more abundant supply of goods. They received this communication with much apparent satisfaction, and jumping out of the boats repeated the speech aloud to their companions. From the general exclamation of "*teyma*," which followed, and from perceiving many of the elderly men retire to a distance, I conceived that they acquiesced in the propriety of the suggestion, and that they were going away, but I was much deceived. They only retired to concert a plan of attack, and returned in a short time shouting some words which Augustus could not make out. We soon, however, discovered their purport, by two of the three chiefs who were on board the Reliance, jumping out, and, with the others who hurried to their assistance, dragging her towards the south shore of the river. Lieutenant Back desired the chief who remained with him to tell them to desist, but he replied by pointing to the beach, and repeating the word *teyma*, *teyma*, with a good-natured smile. He said, however, something to those who were seated in the canoes that were alongside, on which they threw their long knives and arrows into the boat, taking care, in so doing, that the handles and feathered ends were turned towards the crew, as an indication of pacific intentions.

As soon as I perceived the Reliance moving under the efforts of the natives, I directed the Lion's crew to endeavour to follow her, but our boat remained fast until the Esquimaux lent their aid and dragged her after the Reliance. Two of the most powerful men, jumping on board at the same time, seized me by the wrists and forced me to sit between them; and as I shook them loose two or three times, a third Esquimaux took his station in front to catch my arm whenever I attempted to lift my gun, or the broad dagger which hung by my side. The whole way to the shore they kept repeating the word "*teyma*," beating gently on my left breast with their hands, and pressing mine against their breasts. As we neared the beach, two oomiaks, full of women, arrived, and the "*teymas*" and vociferation were redoubled. The Reliance was first brought to the shore, and the Lion close to her a few seconds afterwards. The three men who held me now leaped ashore, and those who had remained in their canoes taking them out of the water, carried them to a little distance. A numerous party then drawing their knives, and stripping themselves to the waist, ran to the Reliance, and having first hauled her as far up as they could, began a regular pillage, handing the articles to the women, who, ranged in a row behind, quickly conveyed them out of sight. Lieutenant Back and his crew strenuously, but good-humouredly, resisted the attack, and rescued many things from their grasp, but they were overpowered by numbers, and had even some difficulty in preserving their arms. One fellow had the audacity to snatch Vivier's knife from his breast, and to cut the buttons from his coat, whilst three stout Esquimaux surrounded Lieutenant Back with uplifted daggers, and were incessant in their demands for whatever attracted their attention, especially for the anchor buttons which he wore on his waistcoat. In this juncture a young chief coming to his aid, drove the assailants away. In their retreat they carried off a writing desk and cloak, which the chief rescued, and then seating himself on Lieutenant Back's knee, he endeavoured to persuade his countrymen to desist by vociferating "*teyma teyma*," and was, indeed, very active in saving whatever he could from their depredations. The Lion had hitherto been beset by smaller numbers, and her crew, by firmly keeping their seats on the cover spread over the cargo, and by beating the natives off with the butt-ends of their muskets, had been able to prevent any article of importance from being carried away. But as soon as I perceived that the work of plunder was going on so actively in the Reliance, I went with Augustus to assist in repressing the tumult; and our bold and active little interpreter rushed among the crowd on shore, and harangued them on their treacherous conduct, until he was actually hoarse. In a short time, however, I was summoned back by Duncan, who called out to me that the Esquimaux had now commenced in earnest to plunder the Lion, and on my return, I found the sides of the boat lined wlth men as thick as they could stand, brandishing their knives in the most furious manner, and attempting to seize every thing that was moveable; whilst another party was ranged on the outside ready to bear away the stolen goods. The Lion's crew still kept their seats, but as it was impossible for so mall a number to keep off such a formidable and determined body, several articles were carried off. Our principal object was to prevent the loss of the arms, oars, or masts, or any thing on which the continuance of the voyage, or our personal safety, depended. Many attempts were made to purloin the box containing the astronomical instruments, and Duncan, after thrice rescuing it from their hands, made it fast to his leg with a cord, determined that they should drag him away also if they took it.

In the whole of this unequal contest, the self-possession of our men was not more conspicuous than the coolness with which the Esquimaux received the heavy blows dealt to them with the butts of the muskets. But at length, irritated at being so often foiled in their attempts, several of them jumped on board and forcibly endeavoured to take the daggers and shot-belts that were about the men's persons; and I myself was engaged with three of them who were trying to disarm me. Lieutenant Back perceiving our situation, and fully appreciating my motives in not coming to extremities, had the kindness to send to my assistance the young chief who had protected him, and who, on his arrival, drove my antagonists out of the boat. I then saw that my crew were nearly overpowered in the fore part of the boat, and hastening to their aid, I fortunately arrived in time to prevent George Wilson from discharging the contents of his musket into the body of an Esquimaux. He had received a provocation of which I was ignorant until the next day, for the fellow had struck at him with a knife, and cut through his coat and waistcoat; and it was only after the affray was over that I learned that Gustavus Aird, the bowman of the Lion, and three of the Reliance's crew, had also narrowly escaped from being wounded, their clothes being cut by the blows made at them with knives. No sooner was the bow clear of one set of marauders, than another party commenced their operations at the stern. My gun was now the object of the struggle, which was beginning to assume a more serious complexion, when the whole of the Esquimaux suddenly fled, and hid themselves behind the drift timber and canoes on the beach. It appears that by the exertions of the crew, the Reliance was again afloat, and Lieutenant Back wisely judging that this was the proper moment for more active interference, directed his men to level their muskets, which had produced that sudden panic. The Lion happily floated soon after, and both were retiring from the beach, when the Esquimaux having recovered from their consternation, put their kaiyacks in the water, and were preparing to follow us; but I desired Augustus to say that I would shoot the first man who came within range of our muskets, which prevented them.

It was now about eight o'clock in the evening, and we had been engaged in this harrassing contest for several hours, yet the only things of importance which they had carried off were the mess canteen and kettles, a tent, a bale containing blankets and shoes, one of the men's bags, and the jib-sails. The other articles they took could well be spared, and they would, in fact, have been distributed amongst them, had they remained quiet. The place to which the boats were dragged is designated by the name of Pillage Point. I cannot sufficiently praise the fortitude and obedience of both the boats' crews in abstaining from the use of their arms. In the first instance I had been influenced by the desire of preventing unnecessary bloodshed, and afterwards, when the critical situation of my party might have well warranted me in employing more decided means for their defence, I still endeavoured to temporize, being convinced that as long as the boats lay aground, and we were beset by such numbers, armed with long knives, bows, arrows, and spears, we could not use fire-arms to advantage. The howling of the women, and the clamour of the men, proved the high excitement to which they had wrought themselves; and I am still of opinion that, mingled as we were with them, the first blood we had shed would have been instantly revenged by the sacrifice of all our lives.

The preceding narrative shows that, bad as the general conduct of the Esquimaux was, we had some active friends amongst them; and I was particularly desirous of cultivating a good understanding with them, for we were as yet ignorant of the state of the ice at sea, and did not know how long we should have to remain in their neighbourhood. I was determined, however, now to keep them at bay, and to convince them, if they made any further attempts to annoy us, that our forbearance had proceeded from good-will, and not from the want of power to punish them. We had not gone above a quarter of a mile from Pillage Point before the boats again took the ground at the distance of one hundred and fifty yards from the shore; and having ascertained by the men wading in every direction, that there was no deeper water, we made the boats fast side by side, and remained in that situation five hours.

Shortly after the boats had been secured, seven or eight of the natives walked along the beach, and carrying on a conversation with Augustus, invited him to a conference on shore. I was at first very unwilling to permit him to go, but the brave little fellow entreated so earnestly that I would suffer him to land and reprove the Esquimaux for their conduct, that I at length consented, and the more readily, on seeing that the young chief who had acted in so friendly a manner was amongst the number on the beach. By the time that Augustus reached the shore, the number of Esquimaux amounted to forty, and we watched with great anxiety the animated conversation he carried on with them. On his return he told us that its purport was as follows:—"Your conduct," said he, "has been very bad, and unlike that of all other Esquimaux. Some of you even stole from me, your countryman, but that I do not mind; I only regret that you should have treated in this violent manner the white people who came solely to do you kindness. My tribe were in the same unhappy state in which you now are, before the white people came to Churchill, but at present they are supplied with every thing they need, and you see that I am well clothed; I get all that I want, and am very comfortable. You cannot expect, after the transactions of this day, that these people will ever bring goods to your country again, unless you show your contrition by returning the stolen goods. The white people love the Esquimaux, and wish to show them the same kindness that they bestow upon the Indians: do not deceive yourselves, and suppose that they are afraid of you; I tell you they are not, and that it is entirely owing to their humanity that many of you were not killed to-day; for they have all guns, with which they can destroy you either when near or at a distance. I also have a gun, and can assure you that if a white man had fallen, I would have been the first to have revenged his death."

The veracity of Augustus was beyond all question with us; such a speech delivered in a circle of forty armed men, was a remarkable instance of personal courage. We could perceive, by the shouts of applause with which they filled the pauses in his harangue, that they assented to his arguments, and he told us that they had expressed great sorrow for having given us so much cause of offence, and pleaded, in mitigation of their conduct, that they had never seen white people before, that every thing in our possession was so new to them, and so desirable, that they could not resist the temptation of stealing, and begged him to assure us that they never would do the like again, for they were anxious to be on terms of friendship with us, that they might partake of the benefits which his tribe derived from their intercourse with the white people. I told Augustus to put their sincerity to the test by desiring them to bring back a large kettle and the tent, which they did, together with some shoes, having sent for them to the island whither they had been conveyed. After this act of restitution, Augustus requested to be permitted to join a dance to which they had invited him, and he was, for upwards of an hour, engaged in dancing and singing with all his might in the midst of a company who were all armed with knives, or bows and arrows. He afterwards told us that he was much delighted on finding that the words of the song, and the different attitudes of the dances, were precisely similar to those used in his own country when a friendly meeting took place with strangers. Augustus now learned from them that there was a regular ebb and flow of the tide in this bay, and that when the sun came round to a particular point there would be water enough to float the boats, if we kept along the western shore. This communication relieved me from much anxiety, for the water was perfectly fresh, and from the flood-tide having passed unperceived whilst we were engaged with the Esquimaux, it appeared to us to have been subsiding for the preceding twelve hours, which naturally excited doubts of our being able to effect a passage to the sea in this direction.

The Esquimaux gradually retired as the night advanced; and when there were only a few remaining, two of our men were sent to a fire which they had made, to prepare chocolate for the refreshment of the party. Up to this period we remained seated in the boats, with our muskets in our hands, and keeping a vigilant look out on Augustus, and the natives around him. As they had foretold, the water began to flow about midnight, and by half past one in the morning of the 8th it was sufficiently deep to allow of our dragging the boats forward to a part where they floated. We pulled along the western shore about six miles, till the appearance of the sky bespoke the immediate approach of a gale; and we had scarcely landed before it came on with

Saturday, 8th.

violence, and attended with so much swell as to compel us to unload the boats and drag them up on the beach.

The whole party having been exhausted by the labour and anxiety of the preceding twenty-four hours, two men were appointed to keep watch, and the rest slept until eleven o'clock in the morning, when we began to repair the damage which the sails and rigging had sustained from the attempts made by the Esquimaux to cut away the copper thimbles. We were thus employed when Lieutenant Back espied, through the haze, the whole body of the Esquimaux paddling towards us. Uncertain of the purport of their visit, and not choosing to open a conference with so large a body in a situation so disadvantageous as our present one, we hastened to launch the boats through the surf, and load them with our utmost speed; conceiving that when once fairly afloat, we could keep any number at bay. We had scarcely pulled into deep water before some of the kaiyacks had arrived within speaking distance, and the man in the headmost one, holding out a kettle, called aloud that he wished to return it, and that the oomiak which was some distance behind, contained the things that had been stolen from us, which they were desirous of restoring, and receiving in return any present that we might be disposed to give. I did not deem it prudent, however, for the sake of the few things in their possession which we required, to hazard their whole party collecting around us, and, therefore, desired Augustus to tell them to go back; but they continued to advance until I fired a ball ahead of the leading canoe, which had the desired effect—the whole party veering round, except four, who followed us for a little way, and then went back to join their companions.

I have been minute in my details of our proceedings with these Esquimaux, for the purpose of elucidating the character of the people we had to deal with; and I feel that the account would be incomplete without the mention, in this place, of some communications made to us in the month of August following, which fully explained the motives of their conduct. We learned that up to the time that the kaiyack was upset, the Esquimaux were actuated by the most friendly feelings towards us, but that the fellow whom we had treated so kindly after the accident, discovering what the boats contained, proposed to the younger men to pillage them. This suggestion was buzzed about, and led to the conference which the old men held together when I desired them to go away, in which the robbery was decided upon, and a pretty general wish was expressed that it should be attended with the total massacre of our party. Providentially a few suggested the impropriety of including Augustus; and for a reason which could scarcely have been imagined. "If we kill him," said they, "no more white people will visit our lands, and we shall lose the opportunity of getting a supply of their valuable goods; but if we spare him, he can be sent back with a story which we shall invent to induce another party of white people to come among us." This argument prevailed at the time; but after the interviews with Augustus at the dance, they retired to their island, where they were so much inflamed by the sight of the valuable articles which they had obtained, that they all, without exception, regretted that they had allowed us to escape. While in this frame of mind the smoke of our fire being discovered, a consultation was immediately held, and a very artful plan laid for the destruction of the party, including Augustus, whom they conceived to be so firmly attached to us that it was in vain to attempt to win him to their cause. They expected to find us on shore; but to provide against the boats getting away if we should have embarked, they caused some kettles to be fastened conspicuously to the leading kaiyack, in order to induce us to stop. The kaiyacks were then to be placed in such a position as to hamper the boats, and their owners were to keep us in play until the whole party had come up, when the attack was to commence. Through the blessing of Providence, their scheme was frustrated.

But to resume the narrative of the voyage. The breeze became moderate and fair; the sails were set, and we passed along the coast in a W.N.W. direction, until eleven in the evening, when we halted on a low island, covered with drift wood, to repair the sails, and to put the boats in proper order for a sea voyage. The continuance and increase of the favourable wind urged us to make all possible despatch, and at three in the morning of the 9th again embarking, we kept in three fathoms water at the distance of two miles from the land. After sailing twelve miles, our progress was completely stopped by the ice adhering to the shore, and stretching beyond the limits of our view to seaward. We could not effect a landing until we had gone back some miles, as we had passed a sheet of ice which was fast to the shore; but at length a convenient spot being found, the boats were hauled up on the beach. We quickly ascended to the top of the bank to look around, and from thence had the mortification to perceive that we had just arrived in time to witness the first rupture of the ice. The only lane of water in the direction of our course was that from which we had been forced to retreat:

Sunday, 9th.

in every other part the sea appeared as firmly frozen as in winter; and even close to our encampment the masses of ice were piled up to the height of thirty feet. Discouraging as was this prospect, we had the consolation to know that our store of provision was sufficiently ample to allow of a few days' detention.

The coast in this part consists of black earth, unmixed with stones of any kind, and its general elevation is from sixty to eighty feet, though in some places it swells into hills of two hundred and fifty feet. A level plain, abounding in small lakes, extends from the top of these banks to the base of a line of hills which lie in front of the Rocky Mountains. The plain was clothed with grass and plants, then in flower, specimens of which were collected. We recognised in the nearest range of the Rocky Mountains, which I have named after my much-esteemed companion Dr. Richardson, the Fitton and the Cupola Mountains, which we had seen from Garry Island at the distance of sixty miles. Few patches of snow were visible on any part of the range.

Having obtained observations for longitude and variation, we retired to bed about eight A.M., but had only just fallen asleep when we were roused by the men on guard calling out that a party of Esquimaux were close to the tents; and, on going out, we found the whole of our party under arms. Three Esquimaux had come upon us unawares, and, in terror at seeing so many strangers, they were on the point of discharging their arrows, when Augustus's voice arrested them, and by explaining the purpose of our arrival, soon calmed their fears. Lieutenant Back and I having made each of them a present, and received in return some arrows, a very amicable conference followed, which was managed by Augustus with equal tact and judgment. It was gratifying to observe our visitors jumping for joy as he pointed out the advantages to be derived from an intercourse with the white people, to whom they were now introduced for the first time. We found that they belonged to a party whose tents were pitched about two miles from us; and as they were very desirous that their friends might also enjoy the gratification of seeing us, they begged that Augustus would return with them to convey the invitation; which request was granted at his desire.

Before their departure, marks being set up on the beach one hundred and fifty yards in front of the tent, and twice that distance from the boats, they were informed that this was the nearest approach which any of their party would be permitted to make; and that at this boundary only would gifts be made, and barter carried on. Augustus was likewise desired to explain to them the destructive power of our guns, and to assure them that every person would be shot who should pass the prescribed limit. This plan was adopted in all succeeding interviews with the Esquimaux. After five hours' absence Augustus returned, accompanied by twenty men and two elderly women, who halted at the boundary. They had come without bows or arrows, by the desire of Augustus, and, following his instruction, each gave Lieutenant Back and myself a hearty shake of the hand. We made presents to every one, of beads, fish-hooks, awls, and trinkets; and that they might have entire confidence in the whole party, our men were furnished with beads to present to them. The men were directed to advance singly, and in such a manner as to prevent the Esquimaux from counting our number, unless they paid the greatest attention, which they were not likely to do while their minds were occupied by a succession of novelties.

Our visitors were soon quite at ease, and we were preparing to question them respecting the coast, and the time of removal of the ice, when Augustus begged that he might put on his gayest dress, and his medals, before the conference began. This was the work of a few seconds; but when he returned, surprise and delight at his altered appearance and numerous ornaments so engaged their minds, that their attention could not be drawn to any other subject for the next half hour. "Ah," said an old man, taking up his medals, "these must have been made by such people as you have been describing, for none that we have seen could do any thing like it;" then taking hold of his coat, he asked "what kind of animal do these skins which you and the chiefs wear belong to? we have none such in our country." The anchor buttons also excited their admiration. At length we managed to gain their attention, and were informed that, as soon as the wind should blow strong from the land, the ice might be expected to remove from the shore, so as to open a passage for boats, and that it would remain in the offing until the reappearance of the stars. "Further to the westward," they continued, "the ice often adheres to the land throughout the summer; and when it does break away, it is carried but a short distance to seaward, and is brought back whenever a strong wind blows on the coast. If there be any channels in these parts, they are unsafe for boats, as the ice is continually tossing about." "We wonder, therefore," they said, "that you are not provided with sledges and dogs, as our men are, to travel along the land, when these interruptions occur." They concluded by warning us not to stay to the westward after the stars could be seen, because the winds would then blow strong from the sea, and pack the ice on the shore. On further inquiry we learned that this party is usually employed, during the summer, in catching whales and seals, in the vicinity of the Mackenzie, and that they seldom travel to the westward beyond a few days' journey. We were, therefore, not much distressed by intelligence which we supposed might have originated in exaggerated accounts received from others. In the evening Augustus returned with them to their tents, and two of the men undertook to fetch a specimen of the rock from Mount Fitton, which was distant about twenty miles. The following observations were obtained:—Latitude 69° 1′ 24″ N.; longitude 137° 35′ W.; variation 46° 41′ E.; dip 82° 22′.

The party assembled at divine service in the evening. The wind blew in violent squalls during the night, which brought such a heavy swell upon the ice, that the larger mases near the encampment were broken before the morning of the 10th, but there was no change in the main body.

Monday, 10th.

The Esquimaux revisited us in the morning, with their women and children; the party consisted of forty-eight persons. They seated themselves as before, in a semicircle, the men being in front, and the women behind. Presents were made to those who had not before received any; and we afterwards purchased several pairs of seal-skin boots, a few pieces of dressed seal-skin, and some deer-skin cut and twisted, to be used as cords. Beads, pins, needles, and ornamental articles, were most in request by the women, to whom the goods principally belonged, but the men were eager to get any thing that was made of iron. They were supplied with hatchets, files, ice chisels, fire-steels, Indian awls, and fish-hooks. They were very anxious to procure knives, but as each was in possession of one, I reserved the few which we had for another occasion. The quarter from whence these knives were obtained, will appear in a subsequent part of the narrative. It was amusing to see the purposes to which they applied the different articles given to them; some of the men danced about with a large cod-fish hook dangling from the nose, others stuck an awl through the same part, and the women immediately decorated their dresses with the ear-rings, thimbles, or whatever trinkets they received. There was in the party a great proportion of elderly persons, who appeared in excellent health, and were very active. The men were stout and robust, and taller than Augustus, or than those seen on the east coast by Captain Parry. Their cheek-bones were less projecting than the representations given of the Esquimaux on the eastern coast, but they had the small eye, and broad nose, which ever distinguish that people. Except the young persons, the whole party were afflicted with sore eyes, arising from exposure to the glare of ice and snow, and two of the old men were nearly blind. They wore the hair on the upper lip and chin; the latter, as well as that on their head, being permitted to grow long, though in some cases a circular spot on the crown of the head was cut bare, like the tonsure of the Roman catholic clergy. Every man had pieces of bone or shells thrust through the septum of his nose; and holes were pierced on each side of the under lip, in which were placed circular pieces of ivory, with a large blue bead in the centre, similar to those represented in the drawings of the natives on the N.W. coast of America, in Kotzebue's Voyage. These ornaments were so much valued, that they declined selling them; and when not rich enough to procure beads or ivory, stones and pieces of bone were substituted. These perforations are made at the age of puberty; and one of the party, who appeared to be about fourteen years old, was pointed out, with delight, by his parents, as having to undergo the operation in the following year. He was a good-looking boy, and we could not fancy his countenance would be much improved by the insertion of the bones or stones, which have the effect of depressing the under lip, and keeping the mouth open.

Their dress consisted of a jacket of rein-deer skin, with a skirt behind and before, and a small hood; breeches of the same material, and boots of seal-skin. Their weapons for the chase were bows and arrows, very neatly made; the latter being headed with bone or iron; and for fishing, spears tipped with bone. They also catch fish with nets and lines. All were armed with knives, which they either keep in their hand, or thrust up the sleeve of their shirt. They had received from the Loucheux Indians some account of the destructive effects of guns. The dress of the women differed from that of the men only in their wearing wide trowsers, and in the size of their hoods, which do not fit close to the head, but are made large, for the purpose of receiving their children. These are ornamented with stripes of different coloured skins, and round the top is fastened a band of wolf's hair, made to stand erect. Their own black hair is very tastefully turned up from behind to the top of the head, and tied by strings of white and blue beads, or cords of white deer-skin. It is divided in front, so as to form on each side a thick tail, to which are appended strings of beads that reach to the waist. The women were from four feet and a half to four and three quarters high, and generally fat. Some of the younger females, and the children, were pretty.

It would appear that the walrus does not visit this part of the coast, as none of these people recognised a sketch of one, which Lieutenant Back drew; but they at once knew the seal and rein-deer. We learned that the polar bear is seldom seen, and only in the autumn; and likewise that there are very few of the brown bears, which we frequently saw on the coast eastward of the Copper-mine River. We had already seen a few white whales, and we understood that they would resort to this part of the coast in greater numbers with the following moon.

The habits of these people were similar, in every respect, to those of the tribes described by Captain Parry, and their dialect differed so little from that used by Augustus, that he had no difficulty in understanding them. He was, therefore, able to give them full particulars relative to the attack made by the other party, and they expressed themselves much hurt at their treacherous conduct. "Those are bad men," they said, "and never fail either to quarrel with us, or steal from us, when we meet. They come, every spring, from the eastern side of the Mackenzie, to fish at the place where you saw them, and return as soon as the ice opens. They are distinguished from us, who live to the westward of the river, by the men being tatoed across the face. Among our tribes the women only are tattoed;" having five or six blue lines drawn perpendicular from the under lip to the chin. The speaker added, "If you are obliged to return by this way, before these people remove, we, with a reinforcement of young men, will be in the vicinity, and will willingly accompany you to assist in repelling any attack." Augustus returned with the Esquimaux to their tents, as there was not the least prospect of our getting forward, though the ice was somewhat broken.

A strong breeze from the westward during the night, contributed, with the swell, to the further reduction of the ice, in front of the encampment; and on the morning of the 11th, the wind changed to the eastward, and removed the pieces a little way off shore, though they were tossing too violently for the boats to proceed. The swell having subsided in the afternoon, we embarked; but at the end of a mile and a half were forced to land again, from the ice being fixed to the shore; and as the wind had now become strong, and was driving the loose pieces on the land, the boats were unloaded and landed on the beach. From the summit of an

Tuesday, 11th.

adjoining hill we perceived an unbroken field of ice to the west, and, consequently, a barrier to our progress.

We encamped on the spot which our Esquimaux friends had left in the morning, to remove in their oomiacks and kaiyacks towards the Mackenzie, where they could set their fishing nets, and catch whales and seals. One of them showed his honesty, by returning some arrows, and a piece of a pemmican bag, that we had left at our last resting-place. The men also joined us here with specimens of rock from Mount Fitton.

The Esquimaux winter residences at this spot were constructed of drift timber, with the roots of the trees upwards, and contained from one to three small apartments, beside a cellar for their stores. There were generally two entrances, north and south, so low as to make it necessary to crawl through them. The only aperture was a hole at the top for the smoke, which, as well as the doorways, could be filled up with a block of snow at pleasure. When covered with snow, and with lamps of fire burning within, these habitations must be extremely warm, though to our ideas rather comfortless. Lofty stages were erected near them for the purpose of receiving their canoes, and bulky articles. A north-east gale came on in the evening, and rolled such a heavy surf on the beach, that twice, during the night, we were obliged to drag the boats and cargoes higher up.

Wednesday, 12th. About three the next morning a heavy rain commenced, and continued, without intermission, through the day; at which we were delighted, however comfortless it made our situation, because we saw the ice gradually loosening from the land under its effects. We found the keeping a tide-pole fixed in the loose gravel beach impracticable here, as well as at the last resting-place, on account of the swell. It appeared to be high water this morning at half past one A.M., and that the rise of tide was about two feet. I need nardly observe that we had the sun constantly above the horizon, were it not for the purpose of mentioning the amusing mistakes which the men made as to the hour. In fact, when not employed, a question as to the time of day never failed to puzzle them, except about midnight, when the sun was near the northern horizon.

Lieutenant Back missing the protractor which he used for laying down his bearings on the map, Augustus set off in the rain early this forenoon to recover it from an Esquimaux woman, whom he had seen pick it up. The rain ceased in the afternoon, the wind gradually abated, and by eight in the evening it was calm. A south wind followed, which opened a passage for the boat, but Augustus was not in sight. At midnight we became greatly alarmed for his safety, having now found that he had taken his gun, which we supposed the natives might have endeavoured to wrest from him, and we were on the point of despatching a party in search of him, when he arrived at four in the morning of the 13th, much fatigued, accompanied by three of the natives. His journey had been lengthened by the Esquimaux having gone farther to the eastward than he had expected, but he had recovered the protractor, which had been kept in their ignorance of its utility to us. His companions brought five white fish, and some specimens of crystal, with other stones, from the mountains, which we purchased, and further rewarded them for their kindness in not allowing Augustus to return alone.

Thursday, 13th.

The boats were immediately launched, and having pulled a short distance from the land, we set the sails, our course being directed to the outer point in view, to avoid the sinuosities of the coast. We passed a wide, though not deep bay, whose points were named after my friends Captains Sabine and P. P. King; and we were drawing near the next projection, when a compact body of ice was discovered, which was joined to the land ahead. At the same time a dense fog came on, that confined our view to a few yards; it was accompanied by a gale from the land, and heavy rain. We had still hopes of getting round the point, and approached the shore in that expectation, but found the ice so closely packed that we could neither advance nor effect a landing. We, therefore, pulled to seaward, and turned the boat's head to the eastward, to trace the outer border of the ice. In this situation we were exposed to great danger from the sudden change of wind to S.E., which raised a heavy swell, and brought down upon us masses of ice of a size that, tossed as they were by the waves, would have injured a ship. We could only catch occasional glimpses of the land through the fog, and were kept in the most anxious suspense, pulling in and out between the floating masses of ice, for five hours, before we could get near the shore. We landed a little to the west of Point Sabine, and only found sufficient space for the boats and tents between the bank and the water. The rain ceased for a short time in the evening, and during this interval, we perceived, from the top of the bank, that the whole space between us and the distant point, as well as the channel by which we had advanced to the westward, were now completely blocked; so that we had good reason to congratulate ourselves on having reached the shore in safety.

CHAPTER IV.

Babbage River—Meet Natives at Herschel Island—Their Trade with the Russians, through the Western Esquimaux—Ascend Mount Conybeare—Boundary of the British Dominions on this Coast—Delayed at Icy Reef—Barter Island—Detention at Foggy Island—Return Reef—Limit of outward Voyage.

Although it rained heavily during the night, and the wind blew strong off the land for some hours, there was no other change in the state of the ice on the morning of the 14th, than that the smaller pieces were driven a short way from the beach. The day was foggy and rainy, but the evening fine. The bank under which we were encamped is of the same earthy kind as that described on the 9th, but rather higher and steeper. It contains much wood coal, similar to that found in the Mackenzie River, and at Garry's Island. The beach and the beds of the rivulets that flow through the ravines, consist of coarse gravel. Specimens of its stones, of the coal, and of the plants in flower, were added to the collection. We saw two marmots, and two rein-deer, which were too wary to allow of our getting within shot of them. Between noon and ten P.M. the loose ice was driving in front of the encampment from the N.W. to S.E., and at the latter hour it stopped. We could not detect any difference in the height of the water, and there was a calm the whole time. A light breeze from S.E. after midnight, brought the masses close to the beach. On the morning of the 15th, having perceived that the ice was loosened from the land near the outer point, to which I have given the name of Kay, after some much esteemed relatives, we embarked, and in the course of a few hours succeeded in reaching it, by passing between the grounded masses of ice. On landing at Point Kay, we observed that our progress must again be stopped by a compact body of ice that was fast to the shore of a deep bay, and extended to our utmost view seaward; and that we could not advance farther than the mouth of a river which discharged its waters just round the point. The boats were, therefore, pulled to its entrance, and we encamped. Former checks had taught us to be patient, and we, therefore, commenced such employments as would best serve to beguile the time, consoling ourselves with the hope that a strong breeze would

Friday, 14th.

Saturday, 15th.

soon spring up from the land and open a passage. Astronomical observations were obtained, the map carried on, and Lieutenant Back sketched the beautiful scenery afforded by a view of the Rocky Mountains, while I was employed in collecting specimens of the plants in flower. The men amused themselves in various ways, and Augustus went to visit an Esquimaux family that were on an island contiguous to our encampment.

We now discovered that the Rocky mountains do not form a continuous chain, but that they run in detached ranges at unequal distances from the coast. The Richardson chain commencing opposite the mouth of the Mackenzie, terminates within view of our present situation. Another range, which I have named in honour of Professor Buckland, begins on the western side of Phillips Bay, and extending to the boundary of our view, is terminated by the Conybeare Mountain.

It gave me great pleasure to affix the name of my friend Mr. Babbage to the river we had discovered, and that of Mr. Phillips, Professor of Painting at the Royal Academy, to the bay into which its waters are emptied. We learned from the Esquimaux that this river, which they call Cook-Keaktok, or Rocky River, descends from a very distant part of the interior, though they are unacquainted with its course beyond the mountains. It appeared to us to flow between the Cupola and Barn mountains of the Richardson chain. There are many banks of gravel near its mouth, but above these obstructions the channel appeared deep, and to be about two miles broad. There were no rocks *in sitû*, or large stones, near the encampment; the rolled pebbles on the beach were sandstone of red and light brown colours, greenstone, and slaty limestone. We gathered a fine specimen of tertiary pitch-coal.

Augustus returned in the evening with a young Esquimaux and his wife, the only residents at the house he had visited. They had now quite recovered the panic into which they had been thrown on our first appearance, which was heightened by their being unable to escape from us owing to the want of a canoe. We made them happy by purchasing the fish they brought, and giving them a few presents; they continued to skip and laugh as long as they staid. The man informed us that judging from the rapid decay of the ice in the few preceding days, we might soon expect it to break from the land, so as to allow of our reaching Herschel Island, which was in view; but he represented the coast to the westward of the island as being low, and so generally beset with ice, that he was of opinion we should have great difficulty in getting along. This couple had been left here to collect fish for the use of their companions, who were to rejoin them for the purpose of killing whales, as soon as the ice should break up; and they told us the black whales would soon come after its rupture took place. It would be interesting to ascertain where the whales retire in the winter, as they require to inhale the air frequently. Those of the white kind make their appearance when there are but small spaces of open water; and we afterwards saw two black whales in a similar situation. One might almost infer from these circumstances that they do not remove very far. Is it probable that they go, at the close of the autumn, to a warmer climate? or can the sea be less closely covered with ice in the high northern latitudes? The situation of our encampment was observed to be, latitude 69° 19′ N.; longitude 138° 10½′ W.; variation 46° 16′ E.; and a rise and fall of nine inches in the water. The wind blew from the west during the night, and drove much ice near the boats; but as the masses took ground a little way from the shore, we were spared the trouble of removing the boats higher up the beach.

Sunday, 16th. We were favoured in the forenoon of the 16th by a strong breeze from the land, which, in the course of a few hours, drove away many of these pieces towards Point Kay, and opened a passage for the boats. We immediately embarked to sail over to the western side of Phillips Bay, concluding, from the motion of the ice, that it must now be detached from that shore. On reaching it, we had the pleasure of finding an open channel close to the beach, although the entrance was barred by a stream of ice lying aground on a reef. The boats being forced by poles over this obstruction, we stood under sail along the coast to about five miles beyond Point Stokes; but there we were again compelled by the closeness of the ice to stop, and from the top of a sand-hill we could not discover any water in the direction of our course. The tents were therefore pitched, and the boats unloaded, and hauled on the beach. Heavy rain came on in the evening, by which we indulged the hope that the ice might be loosened. We were encamped on a low bank of gravel which runs along the base of a chain of sand-hills about one hundred and fifty feet high, and forms the coast line. The bank was covered with drift timber, and is the site of a deserted Esquimaux village. The snow still remaining in the ravines was tinged with light red spots. The night was calm, and the ice remained in the same fixed state until six in the morning of the 17th, when, perceiving the pieces in the offing to be

Monday, 17th.

in motion we launched the boats, and by breaking our way at

first with hatchets, and then forcing with the poles through other streams of ice, we contrived to reach some lanes of water, along which we navigated for four hours. A strong breeze springing up from seaward, caused the ice to close so fast upon the boat, that we were obliged to put again to the shore, and land on a low bank, similar to that on which we had rested the night before. It was intersected, however, by many pools and channels of water, which cut off our communication with the land. As we could not obtain, from our present station, any satisfactory view of the state of the ice to the westward, I despatched Duncan and Augustus to take a survey of it from Point Catton, while Lieutenant Back and I made some astronomical observations. They returned after an absence of two hours, and reported that there was water near Herschel Island, and a channel in the offing that appeared to lead to it. We, therefore, embarked; and by pushing the boats between the masses that lay aground, for some distance, we succeeded in reaching open water at the entrance of the strait which lies between the island and the main, and through which the loose pieces of ice were driving fast to the westward. Having now the benefit of a strong favourable breeze, we were enabled to keep clear of them, and made good progress. Arriving opposite the S.E. end of Herschel Island, we perceived a large herd of rein-deer just taking the water, and on approaching the shore to get within shot, discovered three Esquimaux in pursuit. These men stood gazing at the boats for some minutes, and after a short consultation, we observed them to change the heads of their arrows, and prepare their bows. They then walked along the south shore, parallel to our course, for the purpose, as we soon found, of rejoining their wives. We reached the place at which the ladies were before them, and though invited to land, we were not able, on account of the surf. Augustus was desired to assure them of our friendship, and of our intention to stop at the first sheltered spot, to which they and their husbands might come to receive a present. More than this our liittle friend could not be prevailed upon to communicate, because they were "old wives;" and it was evident that he considered any further conversation with women to be beneath his dignity. On passing round the point we discovered that the ice was closely packed to leeward, and such a heavy swell setting upon it, that it was unsafe to proceed. We, therefore, encamped, and Augustus set off immediately to introduce himself to the Esquimaux. The tents were scarcely pitched, and the sentinels placed, before he returned, accompanied by twelve men and women, each bringing a piece of dried meat, or fish, to present to us. We learned from them that the boats, when at a distance, had been taken for pieces of ice; but when we drew near enough for them to distinguish the crews, and they perceived them clothed differently from any men they had seen, they became alarmed, and made ready their arrows, as we had observed. On receiving some presents, they raised a loud halloo, which brought five or six others from an adjoining island, and in the evening there was a further addition to the party of some young men, who had been hunting, and who afterwards sent their wives to bring us a part of the spoils of their chase. They remained near the tents the greater part of the night, and testified their delight by dancing and singing. An old woman, whose hair was silvered by age, made a prominent figure in these exhibitions.

The information we obtained from them confirmed that which we had received from the last party, namely, that they procure the iron, knives, and beads, through two channels, but principally from a party of Esquimaux who reside a great distance to the westward, and to meet whom they send their young men every spring with furs, seal-skins, and oil, to exchange for those articles; and also from the Indians, who come every year from the interior to trade with them by a river that was directly opposite our encampment; which I have, therefore, named the Mountain Indian River. These Indians leave their families and canoes at two days' march from the mouth of the river, and the men come alone, bringing no more goods than they intend to barter. They were represented to be tall stout men, clothed in deer-skins, and speaking a language very dissimilar to their own. They also said that the Esquimaux to the westward, speak a dialect so different from theirs, that at the first opening of the communication, which was so recent as to be within the memory of two of our present companions, they had great difficulty in understanding them. Several quarrels took place at their first meetings, in consequence of the western party attempting to steal; but latterly there has been a good understanding between them, and the exchanges have been fairly made.

Our visitors did not know from what people either the Indians or the Esquimaux obtained the goods, but they supposed from some "Kabloonacht," (white people,) who reside far to the west. As the articles we saw were not of British manufacture, and were very unlike those sold by the Hudson's Bay Company to the Indians, it cannot be doubted that they are furnished by the Russian Fur Traders, who receive in return for them all the furs collected on this northern coast. Part of the Russian iron-work is conveyed to the Esquimaux dwelling on the coast east of the Mackenzie. The western Esquimaux use tobacco, and some of our visitors had smoked it, but thought the flavour very disagreeable. Until I was aware of their being acquainted with the use of it, I prohibited my men from smoking in their presence, and afterwards from offering their pipes to the Esquimaux at any time. At the conclusion of this conference, our visitors assured us, that having now become acquainted with white people, and being conscious that the trade with them would be beneficial, they would gladly encourage a further intercourse, and do all in their power to prevent future visitors from having such reception as we had on our arrival in these seas. We learned that this island, which has been distinguished by the name of Herschel, is much frequented by the natives at this season of the year, as it abounds with deer, and its surrounding waters afford plenty of fish. It is composed of black earth, rises, in its highest point, to about one hundred feet, and at the time of our visit was covered with verdure. The strait between it and the main shore, is the only place that we had seen, since quitting the Mackenzie, in which a ship could find shelter; but even this channel is much interrupted by shoals. Latitude 69° 33½′ N.; longitude 139° 3′ W.; were observed at the encampment.

Tuesday, 18th.

On the morning of the 18th the fog was so thick that we could not see beyond the beach. It dispersed about noon, and we discovered that there was a channel of open water near the main shore, though in the centre of the strait the ice was heavy, and driving rapidly to the north-west. We embarked at once, in the expectation of being able to penetrate between the drift ice and the land, but the attempt was frustrated by the shallowness of the water; and the fog again spreading as thick as before, we landed on a sand-bank. We were soon visited by another party of the Esquimaux, who brought deer's meat for sale; and although the whole quantity did not amount to a deer, we had to purchase it in small pieces. This practice of dividing the meat among the party, we found to prevail throughout the voyage; and they avowed as their reason for it, the desire that every one might obtain a share of the good things we distributed. One of the men drew on the sand a sketch of the coast to the westward, as far as he was acquainted with it; from which it appeared that there was a line of reefs in front of the coast the whole way; the water being deep on the outside of them, but on the inside too shallow even for their oomiacks to float. We subsequently found that his knowledge of the coast did not extend beyond a few days' march.

The atmosphere becoming more clear about two P.M., we

again embarked, and endeavoured to get to seaward. The boats, however, soon grounded; and finding all our attempts to push through any of the channels between the reefs ineffectual, we pulled back close to Herschel Island. Following, then, the course of the drift ice, we passed near to its south-west point, which was found to be the only deep passage through the strait. We afterwards entered into a fine sheet of open water, the main body of the ice being about half a mile to seaward, and only a few bergs lying aground in the direction of our course. The outer parts of the island appeared closely beset with it. At the end of five miles we discerned another large party of Esquimaux, encamped on a reef; they waved their jackets as signals for us to land, which we declined doing, as we perceived the water to be shallow between us and them. They ran along the beach as far as the end of the reef, tempting us by holding up meat. Only two of the party were provided with canoes, and they followed us to a bluff point of the main shore, on which we landed. These proved to be persons whom we had seen at Herschel Island, and who had visited the Esquimaux in this quarter on purpose to make them acquainted with our arrival. We were happy to learn from them that we should not see any more of their countrymen for some time, because, while surrounded by them, the necessity of closely watching their motions, prevented us from paying due attention to other objects. Resuming our voyage, we pulled along the outer border of a gravel reef, about two hundred yards broad, that runs parallel to, and about half a mile from, the coast, having a line of drift ice on the outside of us. The wind being contrary, and the evening cold, temperature 40°, we encamped on the reef at eight P.M., where we found plenty of drift timber; the water was brackish. The distance travelled this day was eight miles and a half. The main shore opposite the encampment was low to a great distance from the coast; it then appeared to ascend gradually to the base of the Buckland chain of mountains.

Wednesday, 19th. The following morning being calm, and very fine, the boats were launched at three A.M., and we set off in high spirits; but after pulling three miles, we perceived the channel of open water becoming narrow, and the pieces of ice heavier than any we had before seen, some of them being aground in three fathoms water. At six A.M., after having gone five miles and a half, we were stopped by the ice which adhered to the reef, and was unbroken to seaward. Imagining we saw water at some distance beyond this barrier, we were induced to drag the boats across the reef, and launch them into the channel on the inside, in the hope of reaching it. This proved to be a bay, at the head of which we arrived in a short time. It was then discovered that a fog hanging over the ice had been mistaken for water. The boats were, therefore, reconveyed across the reef, the tents pitched, and we had to draw largely on our nearly exhausted stock of patience, as we contemplated the dreary view of this compact icy field. A herd of rein-deer appeared very opportunely to afford some employment, and most of the men were despatched on the chase, but only one was successful. The following observations were obtained:—Latitude 69° 36′ N.; longitude 139° 42′ W.; variation 46° 13′ E. Being now abreast of Mount Conybeare, Lieutenant Back and I were on the point of setting out to visit its summit, when we were stopped by a very dense fog that accompanied a fresh breeze from the N.W., followed by heavy rain. The weather continued bad, until ten the following morning; the ice near the beach was broken into smaller pieces, but as yet too closely packed for our proceeding. The water being brackish in front of the reef, we despatched two men to bring some from the pools at a distance inland, which was found to have the same taste; from this circumstance, as well as from the piles of drift wood, thrown up far from the coast, one may infer that the sea occasionally washes over this low shore. The ice broken off from large masses, and permitted to drain before it was melted, did not furnish us with better water. A couple of pin-tailed ducks were shot, the only pair seen; the black kind were more numerous, but were not fired at, as they are fishing ducks, and, therefore, not good to eat. We also saw a few geese and swans.

Thursday, 20th.

The atmosphere was calm, and perfectly clear, on the morning of the 21st; and as there was not any change in the position of the ice, I visited Mount Conybeare, accompanied by Duncan and Stewart. Though its distance was not more than twelve miles from the coast, the journey proved to be very fatiguing, owing to the swampiness of the ground between the mountain and the sea. We had also the discomfort of being tormented the whole way by myriads of musquitoes. The plain was intersected by a winding river, about forty yards broad, which we forded, and on its western side found a thicket of willows, none of which were above seven inches in circumference, and only five or six feet high. At the foot of the mountain were three parallel platforms, or terraces, whose heights we estimated at fifty, eighty, and one hundred and thirty feet; composed of transition slate,

Friday, 21st.

the stone of the lowest being of the closest texture. We found the task of climbing above the upper terraces difficult, in consequence of the looseness of the stones, which did not afford a firm footing, but after an hour's labour, we succeeded in reaching the top. The mountain is also composed of slate, but so much weathered near the summit, as to appear a mere collection of stones. Its height above the sea we estimated at eight hundred feet. Two or three hardy plants were in flower, at the highest elevation, which we gathered, though they were of the same kind that had been collected in the lower lands; and during the whole march we did not meet with any plant different from the specimens we had already obtained. On arriving at the top of the mountain, we were refreshed by a strong south wind, which we fondly hoped might reach to the coast, and be of service, by driving the ice from the land. This hope, however, lasted only a few minutes; for, on casting our eyes to seaward, there appeared no open water into which it could be moved, except near Herschel Island. The view into the interior possessed the charm of novelty, and attracted particular regard. We commanded a prospect over three ranges of mountains, lying parallel to the Buckland chain, but of less altitude. The view was bounded by a fourth range of high-peaked mountains, for the most part covered with snow. This distant range was afterwards distinguished by the name of the British Chain; and the mountains at its extremities were named in honour of the then Chancellor of the Exchequer, and President of the Board of Trade—the Right Honourable Mr. Robinson, now Lord Goderich, and Mr. Huskisson. When seen from the coast, the mountains of the Buckland chain appeared to form a continuous line, extending from N. W. by N., to S.E. by S.; but from our present situation we discovered that they were seperated from each other by a deep valley, and a rivulet, and that their longest direction was N.N. E. and S.S.W. The same order prevailed in the three ranges behind the Buckland chain; and the highest of their mountains, like Mount Conybeare, were round and naked at the top; the vallies between them were grassy. We erected a pile of stones of sufficient height to be seen from the sea, and deposited underneath it a note, containing the latitude, longitude, and some particulars relative to the Expedition.

Saturday, 22nd. The 22nd was a calm sultry day, the temperature varying between 58° and 63°, and we were tormented by musquitoes. The ice remained very close to the beach. Impatient of our long detention, we gladly availed ourselves, at three in the morning of the 23rd, of a small opening in the ice, to launch the boats, and push them forward as far as we could get them. We thus succeeded in reaching a lane of water, through which we made tolerable progress, though after two hours and a half of exertion, we were gradually hemmed in, and forced again to encamp at the mouth of a small stream westward of Sir Pulteney Malcolm River. We had, however, the satisfaction of finding, by the observations, that we had gained ten miles. Latitude 69° 36′ N.; longitude 140° 12′ W.; variation 45° 6′ E. The temperature of the water at the surface a quarter of a mile from the shore was 40°, that of the air being 49°. The water was two fathoms deep, ten yards from the beach.

Sunday, 23rd.

The coast here was about fifteen feet high; and from the top of the bank a level plain extended to the base of the mountains, which, though very swampy, was covered with verdure. At this place we first found boulder stones, which were deeply seated in the gravel of the beach. They consisted of greenstone, sandstone, and limestone; the first mentioned being the largest, and the last the most numerous. Having seen several fish leaping in the river, a net was set across its mouth, though without success, owing to the meshes being too large. Two men were despatched to examine the state of the ice; and on their return from a walk of several miles, they reported that, with the exception of a small spot close to the beach, it was quite compact. They had observed, about two miles from the encampment, stumps of drift wood fixed in the ground at certain distances, extending from the coast across the plain towards the Rocky Mountains, in the direction of two piles of stones, which were erected on the top of the latter. We were at a loss to conjecture what motive the Esquimaux could have had for taking so much trouble, unless these posts were intended to serve as decoys for the rein-deer. The party assembled at divine service in the evening, as had been our practice every Sunday.

On the morning of the 24th we were able to make a further advance of two miles and three quarters, by forcing the boats between the masses of ice, as far as the debouche of another rivulet, in latitude 69° 36½′ N., and longitude 140° 19½′ W. Under any other circumstance than that of being beset by ice, the beautifully calm and clear weather we then had would have been delightful; but as our hope of being released rested solely on a strong wind, we never ceased to long for its occurrence. A breeze would have been,

Monday, 24th.

at any rate, beneficial in driving away the musquitoes, which were so numerous as to prevent any enjoyment of the open air, and to keep us confined to a tent filled with smoke, the only remedy against their annoyance.

Tuesday, 25th. We were still detained the two following days, and the only things we saw were a grey wolf, some seals, and some ducks. More tedious hours than those passed by us in the present situation, cannot well be imagined. After the astronomical observations had been obtained and worked, the survey brought up, a sketch made of the encampment, and specimens of the plants and stones in the vicinity collected, there was, literally, nothing to do. The anxiety which was inseparable from such an enterprize as ours, at such an advanced period of the season, left but little disposition to read, even if there had been a greater choice of books in our travelling library, and still less composure to invent amusement. Even had the musquitoes been less tormenting, the swampiness of the ground, in which we sank ancle deep at every step, deprived us of the pleasure of walking. A visit to the Rocky Mountains was often talked of, but they were now at a distance of two days' journey, and we dared not to be absent from the boats so long, lest the ice, in its fickle movements, should open for a short time. Notwithstanding the closeness of the ice, we perceived a regular rise and fall of the water, though it amounted only to seven inches, except on the night of the 24th, when the rise was two feet; but the direction of the flood was not yet ascertained. We found a greater proportion of birch-wood, mixed with the drift timber to the westward of the Babbage than we had done before; between the Mackenzie and that river it had been so scarce, that we had to draw upon our store of bark to light the fires. Some lunar observations were obtained in the afternoon of the 25th, and their results assured us that the chronometers were going steadily. At midnight we were visited by a strong S.W. breeze, accompanied by rain, thunder, and lightning. This weather was succeeded by calm, and a fog that continued throughout the next day, and confined our view to a few yards. Temperature from 41° to 43°. On the atmosphere becoming clear about

Wednesday, 26th. nine in the evening of the 26th, we discovered a lane of water, and immediately embarking, we pulled, for an hour, without experiencing much interruption from the ice. A fresh breeze then sprung up from the N.W., which brought with it a very dense fog, and likewise caused the ice to close so fast upon us, that we were compelled to hasten to the shore. We had just landed, when the channel was completely closed. We encamped on the western side of a river about two hundred yards broad, which, at the request of Lieutenant Back, was named after Mr. Backhouse, one of the under Secretaries of State for Foreign Affairs. It appeared that the water that flowed from this channel had caused the opening by which we had travelled from our last resting-place; for beyond it, the ice was closely packed.

Thursday, 27th.

Some heavy rain fell in the night, and the morning of the 27th was foggy; but the sun, about noon, having dispersed the fog, we discovered an open channel about half a mile from the shore. No time was lost in pushing the boats into it. By following its course to the end, and breaking our way through some streams of ice, we were brought, at the end of eight miles, to the mouth of a wide river that flows from the British range of mountains. This being the most westerly river in the British dominions on this coast, and near the line of demarcation between Great Britain and Russia, I named it the Clarence, in honour of His Royal Highness the Lord High Admiral. Under a pile of drift timber which we erected on the most elevated point of the coast near its mouth, was deposited a tin box, containing a royal silver medal, with an account of the proceedings of the Expedition; and the union flag was hoisted under three hearty cheers, the only salute that we could afford. This ceremony did not detain us longer than half an hour; when we launched into a larger space of open water than we had seen since the 9th of the month. This circumstance, together with the appearance of several seals, and the water becoming more salt, created a hope that we should soon enter upon a brisker navigation. But this too sanguine expectation was dispelled in little more than an hour, by a close and heavy field of ice, which obliged us to pull to the shore. The tent was pitched under a steep bank of mud, in latitude 69° 38′ N.; longitude 140° 46′ W. The soundings this day varied from two to ten fathoms; and the temperature of the air from 37° to 45°. The ice having opened near the beach by noon of the 28th, so as to admit the boats, we embarked, to try if we could not advance by thrusting the masses aside with poles. After spending several hours in this labour, and gaining only two miles, further exertion became ineffectual, owing to the ice being closely packed, and many of the pieces from fifteen to twenty feet high, lying aground. We had, however, gained by the removal the comforts of dry ground, and good water, which had been wanting at the last encampment. Among the drift timber on the beach was a pine tree, seven feet and a quarter

Friday, 28th.

in girth, by thirty-six long. We had previously seen several, little inferior in size.. The temperature this day varied from 39° to 48°. We had observed, for the preceding fortnight, that the musquitoes assailed us as soon as the temperature rose to 45°, and that they retired quickly on its descending below that height.

Saturday, 29th. The morning of the 29th opened with heavy rain and fog; the precursors of a strong gale from E.N.E., which brought back the ice we had already passed, and closely packed it along the beach, but we could not perceive that the wind had the slightest effect on the main body at a distance from the shore. This was a very cold, comfortless day, the temperature between 38° and 42°. On Sunday, 30th. the following morning a brilliant sun contributed with the gale to the dispersion of the mist which had, for some days past, overhung the Rocky Mountains, and we had the gratification of seeing, for the first time, the whole length of the British Chain of Mountains, which are more peaked and irregular in their outline, and more picturesque than those of the Buckland Range. The following observations were obtained here:—Latitude 69° 38′ N.; longitude 140° 51′ W.; variation 45° 43′ E.; Dip 83° 27′. In exploring the bed of a rivulet we found several pieces of quartz, containing pyrites of a very bright colour, which so much attracted the attention of the crews, that they spent several hours in examining every stone, expecting to have their labour rewarded by the discovery of some precious metal.

The gale having abated in the evening, we quickly loaded the boats, and pulled them into a lane of water that we had observed about half a mile from the shore. This, however, extended only a short way to the west, and at the end of a mile and a half inclined towards the beach, the ice beyond it being closely packed. Before the boats could be brought to the land, they received several heavy blows in passing through narrow channels, and over tongues of grounded ice. I walked to the extreme point that we had in view from the tent, and was rejoiced by the sight of a large space of water in the direction of our course; but up to the point the ice was still compact, and heavy. On my way I passed another Esquimaux village, where there were marks of recent visitors.

We witnessed the setting of the sun at eleven P.M.; an unwelcome sight, which the gloomy weather had, till then, spared us; for it forced upon our minds the conviction that the favourable season for our operations was fast passing away, though we had, as yet, made so little progress. This was not the only uncomfortable circumstance that attended us this evening. Our friend Augustus was seized with a shivering fit, in consequence of having imprudently rushed, when in full perspiration, into a lake of cold water, to drag out a rein-deer which he had killed. He was unable to walk on coming out of the water, and the consequence would have been more serious had it not been for the kindness of his companion, Wilson, who deprived himself of his flannels and waistcoat to clothe him. On their arrival at the tent, Augustus was put between blankets, and provided with warm chocolate, and the only inconvenience that he felt next morning was pain in his limbs.

Monday, 31st.

We had several showers of rain during the night, with a steady S.W. breeze, and in the morning of the 31st were delighted by perceiving the ice loosening and driving off the land. We were afloat in a few minutes, and enjoyed the novelty of pulling through an uninterrupted channel as far as Point Demarcation, which has been so named from its being situated in longitude 141° W., the boundary between the British and Russian dominions on the northern coast of America. This point seems to be much resorted to by the Esquimaux, as we found here many winter houses, and four large stages. On the latter were deposited several bundles of seal and deer skins, and several pair of snow-shoes. The snow-shoes were netted with cords of deer-skin, and were shaped like those used by the Indians near the Mackenzie. A favourable breeze now sprang up; and having ascertained, by mounting one of the Esquimaux stages, that there was still a channel of open water between a low island and the main shore, we set sail to follow its course. At the end of three miles we found the water gradually to decrease from three fathoms to as many feet, and shortly afterwards the boats repeatedly took the ground. In this situation we were enveloped by a thick fog, which limited our view to a few yards. We, therefore, dragged the boats to the land, until we could see our way; this did not happen before ten in the evening, when it was discovered from the summit of an eminence, about two miles distant, that though the channel was of some extent, it was very shallow, and seemed to be barred by ice to the westward. We also ascertained that it was bounded to the seaward by a long reef. The night proved very stormy, and we were but scantily supplied with drift wood.

August 1st.

Though the morning of the 1st of August commenced with a heavy gale from E.N.E., and very foggy weather, we proceeded to the reef, after much fatigue in dragging the boats over the flats, under the supposition that our best chance of getting forward would be by passing on the outside of it. But there finding heavy ice lying aground, and so closely packed as to preclude the possibility of putting the boats into the water, it was determined to examine the channel by walking along the shore of the reef. An outlet to the sea was discovered, but the channel was so flat that gulls were, in most parts, wading across; and there was, therefore, no other course than to await the separation of the ice from the reef. On the dispersion of the fog in the afternoon, we perceived that some of the masses of ice were from twenty to thirty feet high; and we derived little comfort from beholding, from the top of one of them, an unbroken surface of ice to seaward.

Wednesday, 2nd. The gale blew without the least abatement throughout the night, and until noon of the 2nd, when it terminated in a violent gust, which overthrew the tents. The field of ice was broken in the offing, and the pieces put in motion; and in the evening there appeared a large space of open water, but we could not take advantage of these favourable circumstances, in consequence of the ice still closely besetting the reef. We remarked large heaps of gravel, fifteen feet above the surface of the reef, on the largest iceberg, which must have been caused by the pressure of the ice; and from the top of this berg we had the satisfaction of discovering that a large herd of rein-deer were marching in line towards the opposite side of the channel. Our party was instantly on the alert, and the best hunters were sent in the Reliance in chase of them. The boat grounded about midway across, and the eager sportsmen jumped overboard and hastened to the shore; but such was their want of skill, that only three fawns were killed, out of a herd of three or four hundred. The supply, however, was sufficient for our present use, and the circumstances of the chase afforded amusing conversation for the evening. The astronomical observations place our encampment in latitude 69° 43′ N.; longitude 141° 30′ W. The temperature this day varied from 40° to 42°.

Thursday, 3rd. On the morning of the 3rd a strong breeze set in from the east, which we were rejoiced to find caused a higher flood in the channel than we had yet seen, and the hope of effecting a passage by its course was revived; as the ice was still fast to the reef, and likely to continue so, it was considered better to occupy ourselves in dragging the boats through the mud, than to continue longer in this irksome spot, where the wood was already scarce, and the water indifferent. The boats, therefore, proceeded with four men in each, while the rest of the crew walked along the shore, and rendered assistance wherever it was necessary, to drag them over the shallow parts. After four hours' labour, we reached the eastern part of the bay, which I have had the pleasure of naming after my friend Captain Beaufort, R. N., and which was then covered with ice. We had also the happiness of finding a channel that led to seaward, which enabled us to get on the outside of the reef; but as we pushed as close as we could to the border of the packed ice, our situation, for the next four hours, was attended with no little anxiety. The appearance of the clouds bespoke the return of fog, and we were sailing with a strong breeze through narrow channels, between heavy pieces of drift ice, on the outside of a chain of reefs that stretched across Beaufort Bay, which we knew could not be approached within a mile, owing to the shallowness of the water.

Beyond Point Humphrys, the water being deep close to the coast, we travelled in more security, though the ice was less open than before. We halted to sup on a gravel reef that extends from the main shore to Point Griffin, having run twenty-eight miles, the greatest distance we had made on one day since our departure from the Mackenzie.

A black whale, and several seals, having been seen just before we landed, the water now decidedly salt, and the ice driving with great rapidity to the westward, were circumstances that we hailed with heartfelt joy; as affording the prospect of getting speedily forward, and in the evening we lost sight of Mount Conybeare, which had been visible since the 9th of July. There were several huts on the reef, and one large tent, capable of holding forty persons, which appeared to have been lately occupied, besides eighteen sledges, that we supposed to have been left by the men who had gone from Herschel Island, to exchange their furs with the western Esquimaux. Among the baggage we found a spoon, made out of the musk ox horn, like those used by the Canadian voyagers. At six this evening we passed the termination of the British chain of Mountains, and had now arrived opposite the commencement of another range, which I named after the late Count Romanzoff, Chancellor of the Russian Empire, as a tribute of respect to the memory of that distinguished patron and promoter of discovery and science.

Having taken the precaution of supplying ourselves with fresh water, we quitted the reef, to proceed on our voyage under sail, but shortly afterwards arrived at very heavy ice, apparently packed. We found, however, a narrow passage, and by forcing the boats through it, reached a more open channel, where the oars could be used. This extended along a reef, so that we could pursue our course with safety, being ready to land in the event of the ice drifting upon us. The sun set this evening at half past ten P.M.; and the temperature of the air during its disappearance was 38°. Between the reefs and the low main land the water was entirely free from ice. After passing Point Sir Henry Martin, we were tempted, by the appearance of a bay, to steer within the reefs, as we could then use the sails, and make a more direct course than by winding among the ice. The water proved so shallow that the boats took the ground, at the distance of three miles from the shore, which caused us to alter our plan, and follow the line of drift ice near the border of the pack. The breeze died away; and in proceeding under oars beyond Point Manning, we descried a collection of tents planted on a low island, with many oomiacks, kaiyacks, and dogs around them. The Esquimaux being fast asleep, Augustus was desired to hail them, and after two or three loud calls, a female appeared in a state of nudity; after a few seconds she called out to her husband, who awoke at the first sound of her voice, and shouting out that strangers were close at hand, the whole space between the tents and the water was, in a few minutes, covered with armed, though naked, people. Their consternation on being thus suddenly roused by strangers, of whose existence they had never heard, can be better imagined than described. We drew near the shore, to let Augustus inform them who we were, and of the purpose of our visit; which produced a burst of acclamation, and an immediate invitation to land. This we declined doing, having counted fifty-four grown persons, and knowing that we had not the means of furnishing such a number with the articles they might crave. Besides, it was evident, from their hurried manner, that they were in a state of high excitement, and might then, perhaps, have been disposed to seize upon every thing within their grasp. Four of the kaiyacks being launched, after we had receded to a proper distance from the island, we allowed them to come alongside; and presents were given to the men. We then learned that these were the people who had conveyed the furs, &c., from Herschel Island, and that the exchange with the Esquimaux had been made at the place where they were encamped, only a few days before. They intended to commence their return this day to Herschel Island, where the iron and beads would

Friday, 4th.

be distributed among their relations, according to the furs, &c. they had supplied. The Esquimaux saluted us at parting with many vociferations of *teyma*, and we continued our journey for five miles; at the end of which, the wind setting in strongly against us, we landed at the western part of Barter Island, to refresh the crew. We then found that a rapid tide was running to the eastward, and at eleven the water had risen one foot, from the time of our landing. The tents were scarcely pitched before we saw two kaiyacks coming towards us from the westward, and the man in the headmost accepted, without hesitation, our invitation to land. His companion was asleep, and his canoe was driving with the wind and tide; but when awaked by the voice of Augustus, he also came. These were young men returning from hunting to the tents that we had passed; and being much fatigued, they made but a short stay. The only information collected from them was, that the coast before us was similar to that along which we had been travelling, and that the ice was broken from the shore. The latitude 70° 5′ N.; longitude 143° 55′ W.; variation 45° 36′ E.; were observed.

As soon as the latitude had been obtained, we embarked, favoured by wind and tide, to cross the bay, which has been named in honour of the Marquess Camden. The water was of a seagreen colour, perfectly salt, and from three to five fathoms deep; the temperature 35° at the surface, that of the air, 43°. The day was very clear, and exposed to our view the outline of the Romanzoff chain of Mountains, whose lofty peaks were covered with snow. At the end of ten miles we observed four tents planted on a reef, and several women standing about them, who made many signs for us to land, but the surf was rolling too heavily on the beach. As we proceeded, their husbands were perceived on the main shore, in pursuit of a large herd of rein-deer, which they seemed to be surrounding so as to drive the deer into the water, where they would probably spear them to more advantage.

Continuing along the shore beyond the reef at the distance of two miles from the land, the boats touched the ground several times, which made us conclude we were steering into a bay, though its outline could not be seen. The wind changed at the time to the north, blew strong, and raised a heavy swell, which induced us to haul out to seaward, and we soon afterwards discovered an island, which we just reached under sail. From its summit we perceived a chain of low reefs, extending from its northern point for several miles to the westward, on which the wind was then blowing, and bringing down the drift ice. We were, therefore, compelled to halt, and await more moderate weather. This island, like the projecting points of the main shore, is a mere deposit of earthly mud, covered with verdure, about twenty or twenty-five feet high. There was another island adjoining, which seemed to be a collection of boulder stones; from whence it was named.

The ice appeared closely packed to the seaward; nearer to the island were icebergs aground, and within these, streams of loose pieces driving towards the reefs. In the hurry of embarkation from Barter Island, one of the crew of the Reliance left his gun and ammunition, which we regretted the more, from being apprehensive that an accident might happen to the natives. The circumstance was not known before the boats were a great distance from the island, or we should have put back to have recovered it.

A very thick fog came on in the evening. This weather, however, did not prevent our receiving a visit from two of the natives about midnight, who told Augustus that, having scented the smoke of a fire from the opposite side of the bay, they had come to ascertain who had made it. They were armed with bows and arrows, and advanced towards the tent without any alarm. We found that they had been hunting, with several other men, at the foot of the Romanzoff Mountains, and that they were now going to rejoin their friends at Barter Island, with the fruits of a successful chase. Their knowledge of the coast terminated at this place, which is as far to the westward as any of the party from Herschel Island travel.

The western Esquimaux had parted from them seven nights before, but they supposed that they had not made much progress, as their oomiacks were heavily laden. Those people had informed them that the coast to the westward was low, and fronted by reefs, like that we had already passed; the water also was very shallow; they therefore recommended that we should keep on the outside of every reef. Our visiters had no sooner received their presents than they raised a loud cry, which was intended to bring their friends. On the dispersion of the fog at the time, we discovered an oomiack, filled with people paddling, and some other men wading towards us. It being calm, and the swell having abated, we did not wait for their arrival, but embarked at one in the morning of the 5th, and pursued our course to the westward, keeping on the outside of the reefs. The water, however, was very shallow, even at the distance of two miles, and we were much teased by the boats repeatedly touching the

Saturday, 5th.

ground. This was particularly the case when we arrived opposite to the large river, which was named in honour of the late Mr. Canning, where we found the water perfectly fresh, three miles from the land. The ice being more loose abreast of this river, we pulled out to seaward into deep water. The land was then hidden from our view by the haze, though not more than four miles distant, and our course was directed by the masses of ice lying aground; but at the end of three miles, our further progress was stopped at six A.M., by the ice being closely packed on the outer border of a reef, in latitude 70° 7′ N.; longitude 145° 27′ W.

We perceived, on landing, by the driving of the loose pieces of ice, that the tide was running strongly to the eastward, through the channel we had passed along, and that it continued to do so, until ten this morning, during which time the water was falling. It changed at ten, and the water rose one foot before one P.M. This observation would indicate the flood to come from the eastward, though contrary to what was remarked at Barter Island the day before; but in a sea so closely beset with ice, no accurate observations as to the direction of the tide could be obtained.

The Rocky Mountains either terminated abreast of our present situation, or receded so far to the southward as to be imperceptible from the coast a few miles beyond this reef. The ice being somewhat loosened by the flood tide, we embarked at one P.M., to force the boats through the narrow channels, and in the course of two hours reached Point Brownlow, where we landed, for the purpose of ascertaining whether the ice could be avoided by passing into the bay that then opened to our view, trending to the south. We perceived that this bay was in every part flat, and strewed with stones; and that the only prospect of getting forward was by entering the ice again, and pushing to an island about two miles further to the west, which we reached after receiving several heavy blows in passing through the loose ice at the entrance of the strait, between the central reef and the island, where the pieces were much tossed by the tide.

The view from the south-east part of the island led us, at first, to suppose that we might proceed by keeping close to its south shore; but in making the attempt, the boats repeatedly took the ground, and we were obliged to seek a passage by the north side of the island. At the end of a mile in that direction we were stopped by the ice being unbroken from the shore, and closely packed to seaward. Since the day after our departure from the Mackenzie, when we first came to the ice, we had not witnessed a more unfavourable prospect than that before us. No water was to be seen, either from the tents, or from the different points of the island which we visited, for the purpose of examining into the state of the ice. We were now scantily supplied with fuel; the drift timber being covered by the ice high up the bank, except just where the boat had landed.

In the evening a gale came on from the east, and blew throughout the following day: we vainly hoped this would produce some favourable change; and the water froze in the kettle on the night of the 5th. The position of the encampment was ascertained by observation to be, latitude 70° 11′ N.; longitude 145° 50′ W.; variation 42° 56′ E.; so that notwithstanding the obstructions we had met, an advance of two degrees of longitude had been made in the two preceding days.

Sunday, 6th.

This island received the name of Flaxman, in honour of the late eminent sculptor. It is about four miles long and two broad, and rises, at its highest elevation, about fifty feet. In one of the ravines, where a portion of the bank had been carried away by the disruption of the ice, we perceived that the stratum of loose earth was not more than eighteen inches thick, the lower bed being frozen mud; yet this small quantity of soil, though very swampy, nourished grasses, several of the arctic plants, and some few willows, that were about three inches high. Several boulder stones were scattered on its beach, and also in the channel that separates it from the main shore.

Monday, 7th.

An easterly wind gave place to a calm on the morning of the 7th: and as this change, though it produced no effect in loosening the ice to the north, caused more water to flow into the channel between the island and the main, we succeeded with little difficulty in crossing the flats that had before impeded us. Beyond this bar the water gradually deepened to three fathoms; and a favourable breeze springing up, we set the sail, and steered for the outer point of land in sight. We continued in smooth water until we reached Point Thompson, when, having lost the shelter of the ice which was aground on a tongue of gravel projecting from Flaxman Island, we became exposed to an unpleasant swell.

The Lion was very leaky, in consequence of the blows she had received from the ice; but as we could keep her free by baling, we did not lose the favourable moment by stopping to repair her. Our course was continued past Point Bullen, until we came to an island lying three miles from the shore, which proved to be connected with the main land by a reef. Dazzled by the glare of the sun in our eyes, the surf, which was breaking on this reef, was mistaken for a ripple of the tide; and although the sails were lowered, as a measure of precaution, we were so near before the mistake was discovered, that the strength of the wind drove the Lion aground, by which accident she took in much water. The exertions of the crew soon got her afloat, and both boats were pulled to windward of the island. The sails were then set, but as the wind had by this time increased to a strong gale, they were close reefed. We stood along the coast, looking for a favourable landing place, that we might obtain shelter from an approaching storm which the appearance of the sky indicated, and to repair the damage which the Lion had sustained. At length, some posts that had been erected by the Esquimaux on a point, denoted an approachable part of the coast, and we effected a landing after lightening the boats, by carrying part of the cargo two hundred yards through the water. The main shore to the westward of Flaxman Island is so low that it is not visible at the distance of three miles, with the exception of three small hummocks, which look like islands.

The carpenter had finished the repairs of the boat by midnight, and we were prepared to go forward, but were prevented from moving by a very thick fog, which continued throughout the night, and till eleven on the morning of the 8th. The storm continued violent throughout the day, but the fog cleared away for the space of two hours, and enabled us to perceive that the ice, which in the preceding evening had been at a considerable distance from the land, was now tossing about, in large masses, close to the border of the shallow water. We were also enabled, during the interval of clear weather, to ascertain, by astronomical observations, the latitude 70° 16′ 27″ N.; longitude 147° 38′ W.; and variation 43° 15′ E.

Tuesday, 8th.

The hunters were sent out in pursuit of some deer that were seen, and Augustus killed one. They ascertained, during the chase, that we were on an island, separated from the main shore by a channel, fordable at low water. At this encampment we remarked the first instance of regularity in the tide. It was low water at half past nine on the evening of the 7th, and high water at half past two the following morning; the rise being sixteen inches. An equally regular tide was observed on the 8th, but we could not ascertain the direction of the flood. After sunset the squalls became extremely violent; and

Wednesday, 9th.

until three in the afternoon of the 9th, the fog was so dense that every object more distant than forty yards, was hidden. After that period, a partial clearness of the atmosphere discovered to us the waves more high than the day before, and beating heavily against the weather beach of the island. We rejoiced, however, at seeing a large stream of ice to windward, supposing that its presence there would cause the swell to go down, and that we should be able to proceed as soon as the wind should fall. We employed ourselves in observing the dip with Meyer's and the common needles, as well as the magnetic force. The mean dip was 82° 26′. The temperature of the air this day varied from 38° to 45°. High water took place at fifteen minutes after three P.M., the rise being two feet. The water did not fall so low as yesterday, owing to the wind blowing more across the mouth of the bay.

Thursday, 10th.

On the 10th, the continuance of the gale, and of the fog more opaque, if possible, than before, and more wet, were not only productive of irksome detention, but they prevented us from taking exercise; our walks being confined to a space between the marks which the Esquimaux had put up on two projecting points, whither we went at every glimpse of clearness, to examine into the state of the waves. We witnessed with regret, in these short rambles, the havoc which this dreary weather made amongst the flowers. Many that had been blooming on our arrival, were now lying prostrate and withered. These symptoms of decay could not fail painfully to remind us that the term of our operations was fast approaching; and often, at this time, did every one express a wish that we had some decked vessel, in which the provision could be secured from the injury of salt water, and the crew sheltered when they required rest, that we might quit this shallow coast, and steer at once towards Icy Cape. We designated this dreary place by the name of Foggy Island. As an instance of the illusion occasioned by the fog, I may mention that our hunters sallied forth, on more than one occasion, to fire at what they supposed to be deer, on the bank about one hundred yards from the tents, which, to their surprise, took wing, and proved to be cranes and geese.

Friday, 11th.

The wind changed from east to west in the course of the night, and at eight in the morning of the 11th, the fog dispersed sufficiently to allow of our seeing a point bearing N.W. by W., about three miles and a half distant, which we supposed to be an island. We, therefore, hastened to embark; but before the boats could be dragged so far from the shore that they would float, the fog returned. The wind, however, being light, we resolved to proceed, and steer by compass, to the land that had been seen. Soon after quitting the beach we met with shoals, which forced us to alter the course more to the north; and having made the distance at which we estimated the point to be, and being ignorant which way the coast trended beyond it, we rested for some time upon the oars, in the hope that the fog would clear away, even for a short time, to enable us to shape our course anew; but in vain; all our movements in the bay being impeded by the flats that surrounded us, we were compelled to return to Foggy Island, Scarcely had the men made a fire to dry their clothes, which were thoroughly wet from wading over the flats, than the fog again dispersing, we pushed off once more. On this occasion we arrived abreast of the point whilst the weather continued clear, but found a reef, over which the waves washed, stretching to the north-west, beyond the extent of our view. Just as we began to proceed along the reef, the recurrence of the fog rendered it necessary for us to seek for shelter on the shore; and as we were heartily tired of our late encampment, we endeavoured to find another, but the shoals prevented our reaching any landing-place. We, therefore, retraced our course, though with much reluctance, to Foggy Island, which the men declared to be an enchanted island. Though our wanderings this day did not exceed seven miles, the crews were employed upwards of two hours in dragging the boats through the mud, when the temperature of the water was at 40°, and that of the air 41°. They endured this fatigue with the greatest cheerfulness, though it was evident they suffered very much from the cold; and in the evening we witnessed the ill effects of this kind of labour by finding their legs much swelled and inflamed. The fear of their becoming ill from a frequent repetition of such operations made me resolve not to attempt the passage of these flats again till the weather should be so clear that we might ascertain their extent, and see in what way they might be passed with less risk. Fog is, of all others, the most hazardous state of the atmosphere for navigation in an icy sea, especially when it is accompanied by strong breezes, but particularly so for boats where the shore is unapproachable. If caught by a gale, a heavy swell, or drifting ice, the result must be their wreck, or the throwing the provisions overboard to lighten them, so as to proceed into shoal water. Many large pieces of ice were seen on the borders of the shallow water; and from the lowness of the temperature, we concluded that the main body was at no great distance. We had also passed through a stream of perfectly fresh water, which we supposed was poured out from a large river in the intermediate vicinity, but the fog prevented our seeing its outlet. The atmosphere was equally foggy throughout the night, and all the 12th, although the wind had changed to the east, and blew a strong breeze. Winds from this quarter had been extremely prevalent since the preceding April: but on our former visit to the Polar Sea, they had been of rare occurrence, and confined to the spring months, which we passed at Fort Enterprize. The obstinate continuance of fog forms another material difference between this season and the same period of 1821. We were only detained three times in navigating along the coast that year to the east of the Coppermine River; but on this voyage hardly a day passed after our departure from the Mackenzie that the atmosphere was not, at some time, so foggy as to hide every object more distant than four or five miles. The day that I visited Mount Conybeare, and that spent on Flaxman Island, form the only exceptions to this remark. A question, therefore, suggests itself:—Whence arises this difference? which, I presume, can be best answered by reference to the greater accumulation of ice on this coast, and to the low and very swampy nature of the land. There is a constant exhalation of moisture from the ice and swamps during the summer months, which is, perhaps, prevented from being carried off by the vicinity of the Rocky Mountains, and, therefore, becomes condensed into a fog. The coast to the eastward of the Coppermine River is high and dry, and far less encumbered with ice.

Saturday, 12th.

Some deer appearing near the encampment, a party was despatched in pursuit of them; but having been previously fired at by Augustus, they proved too wary. The exertions of the men were, however, rewarded by the capture of some geese and ducks. The whole of the vegetation had now assumed the autumnal tint.

There was not the least abatement in the wind, or change in the murky atmosphere, throughout the 13th. The party assembled at divine service, and afterwards amused themselves as they could in their tents, which were now so saturated with wet as to be very comfortless abodes; and in order to keep ourselves tolerably warm we were obliged to cover the feet with blankets; our protracted

Sunday, 13th.

stay having caused such a great expenditure of the drift-wood, that we found it necessary to be frugal in its use, and only to light the fire when we wanted to cook the meals. The nights, too, we regretted to find, were lengthening very fast; so that from ten P.M. to two A.M., there was too little light for proceeding in any unknown tract.

Monday, 14th.

The wind this day was moderate, but the fog was more dense, and very wet. Tired, however, of the confinement of the tent, most of the party wandered out in search of amusement, though we could not see one hundred yards; and some partridges, ducks, and geese, were shot.

Tuesday, 15th.

The fog was dispersed at seven in the morning of the 15th, by a north-east gale, which created too great a surf on the beach for us to launch the boats, and the fog returned in the evening. The temperature fell to 35°, and in the course of the night ice was formed on the small pools near the encampment. Augustus set off in the afternoon to cross over to the main shore for the purpose of hunting, and to see whether there were any traces of the western Esquimaux, but he found none, and only saw three rein-deer.

Wednesday, 16th.

The weather again became clear, after the sun rose, on the 16th, and we embarked as soon as the flowing of the tide enabled us to launch the boats, all in the highest spirits at the prospect of escaping from this detestable island. We took advantage of the fair wind, set the sails, and steered to the westward parallel to the coast. We had never more than from three to six feet water, for the first seven miles, until we had passed round the reef that projects from the point we had so often attempted to reach, and which was named Point Anxiety.

Between Point Anxiety and Point Chandos, which is eight miles further to the westward, the land was occasionally seen; but after rounding the latter point we lost sight of it, and steered to the westward across the mouth of Yarborough Inlet, the soundings varying from five feet to five fathoms. The fog returned, and the wind freshening, soon created such a swell upon the flats, that it became necessary to haul further from the land; but the drift ice beginning to close around us, we could no longer proceed with safety, and, therefore, endeavoured to find a landing-place. An attempt was made at Point Herald, and another on the western point of Prudhoe bay, but both were frustrated by the shoalness of the water, and the height of the surf. The increasing violence of the gale, however, and density of the fog, rendering it absolutely necessary for us to obtain some shelter, we stood out to seaward, with the view of making fast to a large piece of ice. In our way we fell among gravelly reefs, and arriving at the same time suddenly in smooth water, we effected a landing on one of them. A temporary dispersion of the fog showed that we were surrounded with banks nearly on a level with the water, and protected to seaward by a large body of ice lying aground. The patch of gravel on which we were encamped, was about five hundred yards in circumference, destitute of water, and with no more drift wood than a few willow branches, sufficient to make one fire.

CHAPTER V.

Commence Return to the Mackenzie—Delayed again at Foggy Island—Ice packed on the Reefs near Beaufort Bay, and on the Coast about Clarence River—Pass the Channels near Herschel Island in a Gale and Fog—A sudden Gale—Escape an Attack which the Mountain Indians meditated—Enter the Mackenzie—Peel River—Arrival at Fort Franklin.

Wednesday, 16th.

The period had now arrived when it was incumbent on me to consider, whether the prospect of our attaining the object of the voyage was sufficiently encouraging to warrant the exposure of the party to daily increasing risk, by continuing on. We were now only half way from the Mackenzie River to Icy Cape; and the chance of reaching the latter, depended on the nature of the coast that was yet unexplored, and the portion of the summer which yet remained for our operations.

I knew, from the descriptions of Cook and Burney, that the shore about Icy Cape resembled that we had already passed, in being flat, and difficult of approach; while the general trending of the coast from the Mackenzie to the west-northwest, nearly in the direction of Icy Cape, combined with the information we had collected from the Esquimaux, led me to conclude that no material change would be found in the intermediate portion.

The preceding narrative shows the difficulties of navigating such a coast, even during the finest part of the summer; if, indeed, any portion of a season which had been marked by a constant succession of fogs and gales could be called fine. No opportunity of advancing had been let slip, after the time of our arrival in the Arctic Sea; and the unwearied zeal and exertion of the crews had been required, for an entire month, to explore the ten degrees of longitude between Herschel Island and our present situation, I had, therefore, no reason to suppose that the ten remaining degrees could be navigated in much less time. The ice, it is true, was more broken up, and the sea around our present encampment was clear; but we had lately seen how readily the drift ice was packed upon the shoals by every breeze of wind blowing towards the land. The summer, bad as it had been, was now nearly at an end, and on this point I had the experience of the former voyage for a guide. At Point Turn-again, two degrees to the south of our present situation, the comparatively warm summer of 1821 was terminated on the 17th of August, by severe storms of wind and snow; and in the space of a fortnight afterwards, winter set in with all its severity. Last year, too, on the 18th and following days of the same month, we had a heavy gale at the mouth of the Mackenzie; and appearances did not indicate that the present season would prove more favourable. The mean temperature of the atmosphere had decreased rapidly since the sun had begun to sink below the horizon, and the thermometer had not lately shown a higher temperature than 37°. Ice, of considerable thickness, formed in the night, and the number of the flocks of geese which were hourly seen pursuing their course to the westward, showed that their autumnal flight had commenced.

While a hope remained of reaching Behring Straits, I looked upon the hazard to which we had, on several occasions, been exposed, of shipwreck on the flats, or on the ice, as inseparable from a voyage of the nature of that which we had undertaken; and if such an accident had occurred, I should have hoped, with a sufficient portion of the summer before me, to conduct my party in safety back to the Mackenzie. But the loss of the boats when we should have been far advanced, and at the end of the season, would have been fatal. The deer hasten from the coast as soon as the snow falls; no Esquimaux had been lately seen, nor any winter-houses, to denote that this part of the coast was much frequented; and if we did meet them under adverse circumstances, we could not, with safety, trust to their assistance for a supply of provision; nor do I believe that, if willing, even they would have been able to support our party for any length of time.

Till our tedious detention at Foggy Island, we had had no doubt of ultimate success; and it was with no ordinary pain that I could now bring myself even to think of relinquishing the great object of my ambition, and of disappointing the flattering confidence that had been reposed in my exertions. But I had higher duties to perform than the gratification of my own feelings; and a mature consideration of all the above matters forced me to the conclusion, that we had reached that point beyond which perseverance would be rashness, and our best efforts must be fruitless. In order to put the reader completely in possession of the motives which would have influenced me, had I been entirely a free agent, I have mentioned them without allusion to the clause in my instructions which directed me to commence my return on the 15th or 20th of August, " if, in consequence of slow progress, or other unforeseen accident, it should remain doubtful whether we should be able to reach Kotzebue's Inlet the same season."

In the evening I communicated my determination to the whole party; they received it with the good feeling that had marked their conduct throughout the voyage, and they assured me of their cheerful acquiescence in any order I should give. The readiness with which they would have prosecuted the voyage, had it been advisable to do so, was the more creditable, because many of them had their legs swelled and inflamed from continually wading in ice-cold water while launching the boats, not only when we accidentally ran on shore, but every time that it was requisite to embark, or to land upon this shallow coast. Nor were these symptoms to be overlooked in coming to a determination; for though no one who knows the resolute disposition of British sailors can be surprised at their more than readiness to proceed, I felt that it was my business to judge of their capability of so doing, and not to allow myself to be seduced by their ardour, however honourable to them, and cheering to me.

Compelled as I was to come to the determination of returning, it is a great satisfaction to me to know, as I now do, that the reasons which induced me to take this step were well-founded. This will appear by the following extract from Captain Beechey's official account of his proceedings in advancing eastward from Icy Cape, with which I have been favoured.

"Mr. Elson, (the master,) after quitting the ship off Icy Cape, on the 18th August, had proceeded along the coast without interruption, until the 22nd of the month, when he arrived off a very low sandy spit, beyond which, to the eastward, the coast formed a bay, with a more easterly trending than that on the west side; but it was so low that it could not be traced far, and became blended with the ice before it reached the horizon. It was found impossible to proceed round the spit, in consequence of the ice being grounded upon it, and extending to the horizon in every direction, except that by which the boat had advanced, and was so compact that no openings were seen in any part of it. This point, which is the most northern part of the continent yet known, lies in latitude, by meridian altitude of the sun, 71° 23′ 39″ N.; and longitude, by several sets of lunar distances, both observed on an iceberg, 156° 21′ W.; and is situated one hundred and twenty miles beyond Icy Cape. Between these two stations, and, indeed, to the southward of the latter, the coast is very flat, abounding in lakes and rivers, which are too shallow to be entered by any thing but a baidar. The greater part of the coast is thickly inhabited by Esquimaux, who have their winter-habitations close to the beach.

"The barge had not been off this point sufficiently long to complete the necessary observations, when the same westerly wind, which had induced me to proceed round Cape Lisburn, brought the ice down upon the coast, and left the boat no retreat. It at the same time occasioned a current along shore to the northward, at the rate of three and four miles per hour. The body of ice took the ground in six and seven fathoms water, but pieces of a lighter draft filled up the space between it and the shore, and, hurried along by the impetuosity of the current, drove the barge ashore, but fortunately without staving her. By the exertions of her officers and crew she was extricated from this perilous situation, and attempts were made to track her along the land wherever openings occurred, in execution of which the greatest fatigue was endured by all her crew. At length all efforts proving ineffectual, and the spaces between the ice and the shore becoming frozen over, it was proposed to abandon the boat, and the crew to make their way along the coast to Kotzebue Sound, before the season should be too far advanced. Preparations were accordingly made; and that the boat might not be irrevocably lost to the ship, it was determined to get her into one of the lakes, and there sink her, that the natives might not break her up, and from which she might be extricated the following summer, should the ship return. During this period of their difficulties they received much assistance from the natives, who, for a little tobacco, put their hands to the tow-rope. Their conduct had, in the first instance, been suspicious; but in the time of their greatest distress, they were well disposed, bringing venison, seal's flesh, oil, &c., and offered up a prayer that the wind would blow off the shore, and liberate the boat from her critical situation. Before the necessary arrangements were made respecting the barge, appearances took a more favourable turn; the ice began to move off shore, and after much tracking, &c., the boat was got clear, and made the best of her way toward the sound; but off Cape Lisburn she met with a gale of wind, which blew in eddies so violently, that it is said the spray was carried up to the tops of the mountains; and the boat, during this trial, behaved so well, that not a moment's anxiety for her safety was entertained. I must not close the account without expressing my warm approbation of the conduct of Mr. Elson."

The barge rejoined Captain Beechey on the 10th September, at Chamisso Island, the Blossom having gone thither to wood and water, and being further forced to quit the coast to the northward, in consequence of strong westerly winds.

Could I have known, or by possibility imagined, that a party from the Blossom had been at the distance of only one hundred and sixty miles from me, no difficulties, dangers, or discouraging circumstances, should have prevailed on me to return; but taking into account the uncertainty of all voyages in a sea obstructed by ice, I had no right to expect that the Blossom had advanced beyond Kotzebue Inlet, or that any party from her had doubled Icy Cape. It is useless now to speculate on the probable result of a proceeding which did not take place; but I may observe, that, had we gone forward as soon as the weather permitted, namely, on the 18th, it is scarcely possible that any change of circumstances could have enabled us to overtake the Blossom's barge.*

Thursday, 17th.

The wind changed to N.E. after midnight, the squalls were more violent, and in the morning of the 17th such a surf was beating on the borders of the reef, that the boats could not be launched. The fog disappeared before the gale about eleven, and during the afternoon we enjoyed the clearest atmosphere that we had witnessed since our departure from Mount Conybeare. This was the first opportunity there had been, for the seven preceding days, of making astronomical observations, and we gladly took advantage of it, to observe the latitude, 70° 26′ N.; longitude 148° 52′ W.; and variation 41° 20′ E. We had likewise the gratification of being able to trace the land round Gwydyr Bay, to its outer point, bearing S. 79 W. ten miles, which I have named after my excellent companion Lieutenant Back, and of seeing a still more westerly hummock, bearing S. 84 W., about fifteen miles, that has been distinguished by the name of my friend Captain Beechey; at which point, in latitude 70° 24′ N., longitude 149° 37′ W., our discoveries terminated. The fog returned at sunset; and as the wind was piercingly cold, and we had neither fire nor room for exercise, we crept between the blankets, as the only means of keeping ourselves warm.

Friday, 18th.

The gale having considerably abated, and the weather being clear, we quitted Return Reef on

* I have recently learned, by letter from Captain Beechey, that the barge turned back on the 25th of August, having been several days beset by the ice. He likewise informs me, that the summer of 1827 was so unfavourable for the navigation of the northern coast of America, that the Blossom did not reach so high a latitude as in the preceding year; nor could his boat get so far to the east of Icy Cape, by one hundred miles. The natives, he says, were numerous, and, in some instances, ill-disposed.

the morning of the 18th, and began to retrace our way towards the Mackenzie. As the waves were still very high to seaward, we attempted to proceed inside of the reefs, but as the boats were constantly taking the ground, we availed ourselves of the first channel that was sufficiently deep to pull on the outside of them. The swell being too great there for the use of the oars, the sails were set double reefed, and the boats beat to the eastward against the wind, between the drift ice and the shallow water.

A gale rose after noon from N.E. by N., which enabled us to shape a course for Foggy Island, where we arrived at three P.M., just at a time when the violence of the squalls, and the increased height of the swell, would have rendered further proceeding very hazardous. We now enjoyed the comforts of a good fire and a warm meal, which we had not had since the evening of the 16th. The men were afterwards employed in erecting a square pile of drift timber, on the highest part of the island fronting the sea, on which a red cornet flag was left flying, and underneath it was deposited, in a tin case, a letter for Captain Parry, containing an account of our proceedings; also a silver medal and a halfpenny: and in order that government might have some chance of hearing of our proceedings, should any accident subsequently befal the party, there was also deposited an unsealed letter, wrapped in bark, addressed to the Russian Fur Traders, in the expectation that the Esquimaux might probably convey it to their Establishment. An ice chissel, a knife, a file, and a hatchet, were hung up on the pile, for the Esquimaux. On digging to erect these posts, the ground was found frozen at the depth of sixteen inches; and the thermometer, during the day, seldom rose above 37°. This evening the temperature was 33°. We were vexatiously detained the

Saturday, 19th.

19th, and following day, by the continuance of the gale, and a thick fog; during which time many large flocks of geese were observed passing away to the westward. The tides were now much higher than during our first visit.

Monday, 21st.

The breeze was moderate on the morning of the 21st, yet we were prevented from embarking until ten o'clock, by the return of the fog. We then hastened to escape from this ill-omened island. The boats were pulled to seaward, so as to gain a sufficient offing for them to pass on the outside of the shallow water; and by the aid of the oars and sails we made good progress, and encamped within sight of Flaxman Island. A black whale, a seal of the largest kind, and numerous flocks of geese were seen in the course of this day. Several stars were visible after ten P.M. Showers of snow fell during the night, but the morning of the 22nd was calm and clear. We embarked at daylight, and in the course of three hours arrived abreast of the east end of Flaxman Island. The ice had broken from the northern shore during our absence, and was now lying about a mile from the land, apparently aground on reefs, as we had observed it to be along the outer border of the one at the west end of the island. The water was much deeper between Flaxman Island and the main, that when we passed in the early part of August. Eastward of Point Brownlow there was an open channel of three or four miles wide. And by keeping close to the borders of the drift ice we avoided the shallows at the mouth of the Canning River, and arrived at Boulder Island about noon. Here we found an Esquimaux grave, containing three bodies, covered with drift timber, and by their side there were placed the canoes, arrows, and fishing implements of the deceased. Not being able to procure fresh water here, we set forward to cross Camden Bay, touched at one of the points to fill the water-casks, and reached Barter Island after dark; the crews much fatigued, having been pulling for nineteen hours. We regretted to find the Esquimaux had visited this spot during our absence, and carried away the gun and ammunition which had been left by mistake at the encampment on the 4th of August, because we were not only apprehensive that some accident might have occurred in the attempt to discharge the gun, but were desirous to prevent the introduction of fire-arms among these people. Being now near the point of the coast at which we had seen a considerable number of the natives, we remained at the encampment until ten o'clock on the morning of the 23d, to clean the guns and issue a fresh supply of ammunition to the party. The day was calm and cloudless; the whole range of the Romanzoff Mountains was in sight, and they appeared to be more covered with snow than when we passed to the westward. A few musquitoes made their appearance, but they were very feeble. Having landed at Point Manning to replenish the water-casks, we afterwards pulled throughout the day close to the edge of the ice, which was still heavy, though loose, and encamped near Point Griffin. Some large-sized medusæ, and several of the gelatinous substances known to seamen by the name of blubber, were found on the beach, which accounted for the number of black ducks that had been seen in the course of the day, as they feed on those substances. The temperature

Tuesday, 22nd.

Wednesday, 23rd.

varied this day from 35° to 46°; and the thermometer rose to 64° at two P.M., when exposed to the sun's rays.

Thursday, 24th. The morning of the 24th was calm; we set forward at daylight, and having proceeded a few miles between heavy floating ice, about half a mile from the shore we met with a large sheet of bay ice of last night's formation, of sufficient thickness to impede though not to stop the boats. Having arrived abreast of Point Humphreys, we steered out to seaward, for the purpose of avoiding the shallows that extend across Beaufort Bay, intending to direct our course in a line for Mount Conybeare, which was in sight. We were then exposed to a long rolling swell, and we soon afterwards perceived that it had driven the ice upon the reefs at the eastern extremity of the bay, which would have precluded our retreat to the shore in the event of the wind rising. It therefore became necessary to penetrate into the pack, and keep by the side of the reefs; but in doing so, the boats were exposed to no little danger of being broken in passing through the narrow channels between the masses of ice which were tossing with the swell, and from which large pieces frequently fell. At six P.M. we passed our former encampment on Icy Reef, and afterwards proceeded through an open space to Demarcation Point, where we encamped, and hauled up the boats to prevent them from being injured by the surf. We found here two families of Esquimaux, which belonged to the party that had been to Barter Island, waiting the return of a man from hunting, in order to follow their companions to the eastward. They showed much joy at seeing us again, and remained the greater part of the night talking with Augustus. The most active young man of the party, not thinking himself sufficiently smart for the occasion, retired to the oomiack to change his dress and mouth ornaments, capering about on his return, evidently proud of his gayer appearance.

Friday, 25th. The morning was foggy, but there being little wind, we launched the boats, and pulled for an hour close to the shore, when we came to a body of ice so closely packed as scarcely to afford a passage, and it was with difficulty that we arrived at Clarence River. There we perceived four tents; near which we had been warned by our visitors last night not to land, as the party had recently lost their parents, and it was feared that, in the state of mind in which they then were, they might be disposed to do us some injury. We pulled near enough to inquire about the gun, and learned that the person who had it was farther to the eastward. The difficulties of forcing a passage were not diminished beyond this place, and we were further impeded in our advance by new ice formed between the larger masses, which required additional labour to break through. The fog cleared away at ten; we halted to breakfast at Backhouse River, and remained whilst Augustus went in pursuit of two rein-deer, one of which he killed.

Renewing our course, we passed on the outside of the ice until we were nearly abreast of Mount Conybeare, when the wind came strong from the eastward, and obliged us to have recourse again to its shelter. This barrier, however, terminated at the end of five miles, and being then exposed to the wind and swell, against which the men were unable to pull, we encamped.

The experience we had now gained of the ice being packed upon this shore by a wind from the sea, assured us of the correctness of the report which the Esquimaux had given, and likewise afforded a reason for their expression of surprise at our being unprovided with sledges, as it was evident, unless a strong wind blew from the land, that the new ice would soon unite the pack with the shore, and preclude the possibility of making the passage in boats, unless by going outside of the ice, which would be extremely hazardous, from the want of shelter in the event of a gale springing up. The pieces of ice were generally from ten to fifteen feet in height, many of them were from twenty to thirty feet: their length was from twenty to one hundred yards. We saw several white whales in the open water, and a flock of white geese at the encampment, which were the first noticed on this coast. The rising of the wind from E.N.E. this afternoon was accompanied by an increase of temperature from 43° to 53°, and we felt a comfortable sensation of warmth, to which we had been strangers for the preceding month.

Saturday, 26th.

We took advantage of a favourable breeze to embark before daylight on the morning of the 26th; at sunrise it increased to a gale, and raised a heavy sea. In two hours we ran to the commencement of the intricate channels leading to Herschel Island, where the Esquimaux seen at Barter Island were encamped on a reef, and apparently gazing in astonishment at the speed of our boats. They made many signs for us to land, which we were desirous of doing had it been practicable for the surf. That the boats might be perfectly manageable, we took two reefs in the sails, and shaped the course for Herschel Island; but scarcely were the sails reset before a fog came on that hid every mark that could guide us; a heavy swell was rolling at the time, and to arrive at the island we had to pass through a channel only about two hundred yards broad. To find this, surrounded as it was by shoals, in the midst of a dense fog, was a task of considerable anxiety and danger, and our situation was not rendered more agreeable by being assailed the whole way with continued shouting from persons to us invisible; our arrival having been communicated by the Esquimaux who first descried us, to their companions on the neighbouring reefs. We effected it, however, and landed in safety, though we did not discover the island till we were within forty yards of its shore. We had scarcely landed before the fog dispersed, and discovered to us a solitary tent on an adjacent point. Three men soon paid us a visit, whom we had not seen before, and they informed us that nearly the whole of the tribe was now collected in the vicinity for the purpose of hunting deer, and catching whales and seals for the winter's consumption. We quitted the island at ten A.M., and steered directly for Point Kay, to avoid the sinuosities of the coast, and the frequent interruption of the Esquimaux, whose tents were observed to be scattered on the beach nearly the whole way to Babbage River. Three men and some women came off to bring us fish, and being liberally rewarded, they went away perfectly happy, singing the praises of the white people. We passed round Point Kay at four P.M., with a moderate breeze from W.N.W., and steered for Point King, keeping about two miles from the land. As the afternoon wore away, gloomy clouds gathered in the north-west; and at six a violent squall came from that quarter, attended with snow and sleet. The gale increased with rapidity: in less than ten minutes the sea was white with foam, and such waves were raised as I had never before been exposed to in a boat. The spray and sea broke over us incessantly, and it was with difficulty that we could keep free by baling. Our little vessels went through the water with great velocity under a close-reefed sail, hoisted about three feet up the main-mast, and proved themselves to be very buoyant. Their small size, however, and the nature of their construction, necessarily adapted for the navigation of shallow rivers, unfitting them for withstanding the sea then running, we were in imminent danger of foundering. I therefore resolved on making for the shore, as the only means of saving the party, although I was aware that, in so doing, I incurred the hazard of staving the boats, there being few places on this part of the coast where there was sufficient beach under the broken cliffs. The wind blowing along the land we could not venture on exposing the boat's side to the sea by hauling directly in, but, edging away with the wind on the quarter, we most providentially took the ground in a favourable spot. The boats were instantly filled with the surf, but they were unloaded and dragged up without having sustained any material damage. Impressed with a sense of gratitude for the signal deliverance we had experienced on this and other occasions, we assembled in the evening to offer up praise and thanksgiving to the Almighty.

Sunday, 27th.

On the 27th the weather was calm; but as a heavy surf prevented our embarkation, we took advantage of the delay to dry our bedding, clothes, and pemmican. The guns were likewise cleaned, and every thing put in order. There as an Esquimaux party at this spot, which had witnessed the landing of the boats in the storm with astonishment, having expected to see every man drowned. Augustus passed the night at their tents; and having brought the whole party to our encampment, the women, with much good nature, sewed soles of seal-skins to the men's mocassins, in order to fit them better for the operation of tracking, in which they were soon to be employed. These Esquimaux had recently returned from a visit to the gang that had pillaged the boats at the mouth of the Mackenzie; and we now learned the intention that had been entertained of destroying our party, along with the other particulars that have been already mentioned.

Monday, 28th.

Our approach to the Mackenzie was marked by the quantity of drift timber floating about. We passed several families of the natives, without visiting them, until we perceived one party taking some fish from their nets, which tempted us to land. The fish were large *tittameg* and *inconnû*, and proved remarkably fine. We again embarked, but having to pull head to the sea, we took in much water, and were glad to seek shelter on a gravel reef, where three Esquimaux tents were pitched. The whole party quickly mustered around us, and we were not a little surprised to find so many inhabitants as twenty-seven, including women and children, in three tents only; but on inquiry we found that the number was not greater than usual. Two of the men were very aged and feeble, the rest were young and active. They practise jumping, as an amusement, from their youth; and we had an opportunity of witnessing some of their feats, which displayed much agility. The women cheerfully repaired our mocassins, and their industry, as well as the good conduct of the men, were rewarded by some valuable presents. We were astonished to learn that there had been fog only a day and a half in this neighbourhood since we passed, but the wind had been generally strong. Augustus gained some information respecting the western Esquimaux, and the coast to the westward, which he did not communicate to me until some days afterwards, otherwise I should have endeavoured to elicit more satisfactory details. It was to the following purport:—The western Esquimaux having purchased the furs from those men that dwell near the Mackenzie, at Barter Island, proceed to the westward again without delay. A few days journey beyond a part of the coast which Augustus understood from description to be Return Reef, the sea is still more shallow than that which we had navigated, and the water is still, except at certain periods of the year, when it is agitated like a strong rapid, by the efflux of the waters of a deep inlet, or strait. The land is visible on both sides from the middle of this opening; the Esquimaux make for the west side, and on reaching it relinquish their canoes, and drag their furs overland to the Russian establishments, which are situated in the interior, where the land is more elevated than on the coast. The Mountain Indians come down annually in large parties to this inlet, and warm contests often arise between them and the Esquimaux. The latter are frequently worsted, from their inferior numbers, and lose their property, which the Indians bring by land to the neighbourhood of Herschel Island, to dispose of to the Esquimaux in that quarter. The direction of the inlet was supposed, by Augustus, to be about south-west. I am inclined to think that it is the estuary of a large river, flowing to the west of the Rocky Mountains, obstructed by sand-banks, like the mouth of the Mackenzie. In the course of the day three Esquimaux, who had seen our tent from a distance, came to visit us. One of them was recognised to have been of the party which attacked us at the mouth of the Mackenzie. He gave Augustus a detailed account of their schemes on that occasion, which exactly corresponded with that we had received on the preceding day. He further told us that the party which had assailed us had certainly removed to the eastward; but if any of them should have remained, to watch our motions, they could be avoided by entering the river by a more westerly branch than the one which we had descended, and offered to guide us thither. This man was very intelligent, and having carefully examined the boats, intimated that he would construct an oomiack after the same plan. We embarked at four in the evening with our new friend for a guide, and in a short time arrived at the main shore where his tent stood, and where he asked the party to encamp, as he intended to go no farther. We were not, however, so disposed; and having filled the casks with fresh water, and made some presents to the women, we pushed off to take advantage of the remaining daylight in getting round a reef which projected far to seaward. We could not effect this, and at sunset, not being able to land on the reef on account of the shallowness of the water, we put back to within a mile and a half of the Esquimaux tents. Garry Island was seen soon after sunset; and the aurora borealis appeared in the night for the first time this season. The temperature of the air varied from 30° to 49°, and that of the sea water was 37° 2′, a quarter of a mile from the shore. A gale coming on in the night, and continuing till the following evening, detained us on shore. During our stay we were visited by a numerous party of Esquimaux, and found it necessary to draw a line round the tents, which they were not permitted to pass. These people told us that Dr. Richardson's party had been seen clear of the Mackenzie, and had given kettles to men in three canoes, after escaping an attempt made by the Esquimaux to drag the boats on shore. This account, showing that the propensity to plunder was not confined to the Esquimaux with whom we had met, excited painful apprehensions for the safety of the eastern party, if they should find it necessary to return by the Mackenzie, because we now learned that the natives collect in numbers near its mouth at the close of summer. In ordinary seasons the weather is mild, and the winds variable until the ice breaks up, which is usually about the end of August, when north-west winds, and stormy weather, are expected. In this season, however, the winds had been so boisterous that the Esquimaux had seldom been able to venture out to sea, and their whale fishery had consequently failed. Our visitors left us about two P.M.; but, shortly afterwards, we heard loud cries, and on looking round saw two young Esquimaux running in breathless haste to announce that a large party of Indians had come down from the mountains with the express purpose of attacking the boats and killing every man of the party. They desired us to embark instantly, as the only means of escape; for the Indians, they said, were already at the tents within our view, and when they left them they were on the point of spreading round us to commence the onset. They further said, that the Indians, having been provoked by our trading with the Esquimaux, had been along the coast in search of us, and that it was only this afternoon they had espied our tents, which, by the fluttering in the wind, they knew did not belong to the Esquimaux. On this discovery they had come to the nearest party of Esquimaux to make known their intention, and to request their aid. They were met by our two young friends, who were out hunting, but who returned with them to their tents, and after learning the

Tuesday, 29th.

plans in agitation, had stolen off to apprize us of our danger. As soon as Spinks returned, who had gone to shoot, we shoved off; and never were men more delighted than our two Esquimaux friends seemed to be at our escape; and especially at that of Augustus, to save whom, they asserted more than once, was their principal motive in coming to us. While Spinks was out of sight, they climbed up to the top of an old house to look for him, with the greatest apparent solicitude, and were the first to discover him returning. Up to the time of his arrival they kept repeating every particular respecting the Indians, and pointing out the mode of avoiding them. It was their intention, they said, to pursue us to the Mackenzie, but that we should get there before them, because there were two rivers in the way which the Indians would have great difficulty in crossing, being unprovided with canoes. They urged us to make all speed, and not to halt in the night, nor to go to sleep; but, if the crew became tired, to put up on an island out of gun-shot of the main shore, because the Indians were armed with guns as well as bows. They instructed Augustus minutely as to the course we were to steer round the reef, and directed us to keep along the main shore until we should come to a large opening, which was the western outlet of the Mackenzie, and had a deep channel. We rewarded their friendly conduct by a considerable present of iron, which they received with an indifference that showed them not to have been actuated by interested motives in making the communications. Previous to the arrival of these men we had perceived the smoke of a distant fire, which we had little regarded, supposing it to have been made by some Esquimaux who were hunting, but which, it seems, was the fire of the Indians. Having pulled round the reef, and being aided by a westerly breeze, we soon regained the main shore, and passed the mouth of the two rivers of which the Esquimaux had spoken. The night beginning to close in we pulled up to the head of an inlet; when heavy rain and squalls coming on, we determined to halt. As soon as the day dawned, which was about half-past two in the morning, we returned to Shoal Water Bay; and, sailing along the coast for two or three miles to the eastward, arrived at another opening, in which the water was fresh, and we did not doubt but it would prove the deep channel by which we had been instructed to ascend. There was plenty of water near its mouth, but it gradually shoaled; and, at the distance of four miles, we ascertained that this promising opening was likewise an inlet. I now relinquished the search for a more westerly outlet than the one by which we had descended, and, therefore, steered for Pillage Point, which soon

Wednesday, 30th.

afterwards came in sight. After dragging the boats for two hours, over the shoals, we rounded Pillage Point at ten A.M., and reached the deep water most opportunely; for, almost at the instant, a violent north-west gale came on, attended by thunder, lightning, and torrents of rain. The wind, however, was fair, and brought so much water into the channel of the river, that we passed, without obstruction, the shallow parts above Pillage Point. A temporary cessation of the rain at noon enabled us to land to breakfast; and we afterwards continued to scud before the gale until sunset, when we encamped. The temperature fell from 48° to 40° in the gale, and we had several showers of snow.

During the above run Augustus entertained us with an account, which he had learned from the two Esquimaux, respecting the Mountain Indians; the substance was as follows: —Seven men of that tribe had been to Herschel Island to trade with the Esquimaux, who showed them the different articles they had received from us, and informed them of our being still on the coast, and that our return by this route was not improbable. This intelligence they set off at once to communicate to the rest of their tribe, who, supposing that we should ruin their trade with the Esquimaux, resolved on coming down in a body to destroy us; and that they might travel with expedition, their wives and families were left behind. They came to the sea coast by the Mountain Indian River, opposite Herschel Island, and finding that we had not returned, but supposing it possible that we might pass them there, as they had no canoes to intercept us, they determined on travelling to the mouth of the Mackenzie, where they could conveniently subsist by fishing and hunting until our arrival. They had been informed of the manner in which we had been robbed by the Esquimaux at that place, and they formed a similar plan of operations. When our crews were wading and launching the boats over the flats in Shoal Water Bay, a few of them were to have offered their assistance, which they imagined would be readily accepted, as we should probably take them for Indians belonging to the Loucheux tribe, with whom we were acquainted. While pretending to aid us they were to have watched an opportunity of staving the boats, so as to prevent them from floating in the deeper channel, which runs close to the land near Pillage Point. The rest of the party, on a signal being given, were then to rush forth from their concealment, and join in the assault. They were, in pursuance of this plan, travelling towards the Mackenzie, when they discovered our tents; and it appeared that the two young men who brought us the intelligence, had been sent as an act of gratitude by an old Esquimaux, to whom we had given a knife and some other things, on the preceding day. After hearing the plans of the Indians, he called the young men aside and said to them, "These white people have been kind to us, and they are few in number, why should we suffer them to be killed? you are active young men, run and tell them to depart instantly. The messengers suggested that we had guns, and could defend ourselves. "True," said he, "against a small force, but not against so large a body of Indians as this, who are likewise armed with guns, and who will crawl under cover of the drift timber, so as to surround them before they are aware; run, therefore, and tell them not to lose a moment in getting away, and to be careful to avoid the flats at the mouth of the river by entering the western channel."

As the goods which the Mountain Indians exchange with the Esquimaux at Herschel Island, are very unlike those issued from any of the Hudson's Bay Company's posts, I conclude that they obtain them from the Russians; but the traders of that nation being prohibited by their government from supplying guns to any Indians, I am at a loss to account for these people having them;—perhaps, the prohibition only applies to the Esquimaux, or the people on the sea coast. That the Mountain Indians have fire-arms we learned, not only on the present occasion, but in our first interview with the Esquimaux, at Herschel Island.

The few general remarks which I have to offer, on the subject of a North-West Passage, will appear in a subsequent part of the narrative; and here I shall only state, that we traced the coast, westward from the mouth of the Mackenzie, three hundred and seventy-four miles, without having found one harbour in which a ship could find shelter.

Thursday, 31st. On the 31st, we continued the ascent of the river, and encamped in the evening within the limit of the spruce fir trees.

September 1st. Favoured by a strong north-west gale, on the 1st of September, we sailed the whole day along the western main shore, and, generally, within view of the Rocky Mountains. One of the numerous bends of the river took us within eight miles of part of the mountains, which appeared to be composed of a yellow stone, and was from eight hundred to a thousand feet in height. In the course of the day we came to the most northerly poplars, where the foliage had now assumed the yellow autumnal hue. The gale continued with strong squalls on the 2nd, and we advanced rapidly under double-reefed sails, though the course of the river was very winding. The temperature of the air varied from 41° to 35°. On the third we had calm weather, and still keeping the western land aboard, we were led into a river which we had not discovered in our descent. The course of this river, was, for a time, parallel to our route, and we took it at first for one of the channels of the Mackenzie; but, in the afternoon, we saw a mountain to the eastward, and ascertained that we were to the Southward of Point Separation. We, therefore, began to descend the river again, and encamped shortly after sunset. Just after it became dark, voices were heard on the opposite side of the river, to which we replied, and soon afterwards, three Indians were observed crossing towards us in canoes. They approached cautiously, but on being invited to land, they did so, though one of them was so great a cripple as to require being carried from the canoe to the fire-side. The alarm these poor people had felt, was soon dissipated by kind treatment. They were armed with bows and arrows only, and clothed in hair skins and leather. Their trowsers were similar to those worn by the lower Loucheux, to which tribe they, probably, belonged. We could communicate with them only by signs, except by using a few words of Chipewyan, which one of them appeared to understand. We collected from them that they knew of Fort Good Hope, but none of them seemed to have visited it, as they had not a single article of European manufacture about their persons. They delineated on a stone the course of the Mackenzie, and of the river we had newly discovered, which appears to flow from the Rocky Mountains, and to break through the same ridge of hill that the Mackenzie does at the Narrows. It is probable, that it was to this river the Loucheux alluded, when they told Sir Alexander Mackenzie, opposite the present site of Fort Good Hope, that there was a river which conducted them to the sea in five days. I have distinguished this river by the name of Peel, in honour of His Majesty's Secretary of State for the Home Department. It is from a quarter to half a mile wide, and its banks are clothed with spruce, birch, and poplar trees, like those of the Mackenzie in the same parallel. We set forward at four A.M. on the 4th, with a strong favourable breeze, and in an hour, passed another river descending from the Rocky Mountains, and nearly as large as the Peel, into which it flows. We regained the Mackenzie at noon, and at five P.M. arrived at Point Separa-

Saturday, 2nd.

Sunday, 3rd.

Monday, 4th.

tion, where we encamped. Here we found the boat, rope, and kettle, in the same state in which we had deposited them. The kettle was a great acquisition to us, because we had suffered much inconvenience in having only one for cooking, after the Esquimaux had robbed us of the others. The temperature varied during the day from 29° to 55°, and, in the evening, the sand flies were troublesome. We quitted our encampment at day light on the 5th, and crossed the river to look for a mark which Dr. Richardson was to have erected, if he returned by the Mackenzie; but not finding any, we deposited a letter and a bag of pemmican, in case he should come at a later period, and that his party should be in want of provision. In the vicinity of the Red River, we met Barbue, the Chief of the Loucheux, and two or three families, who seemed in a sorry condition from want of food, the water being too low for fishing. The chief appeared very anxious to communicate some intelligence, which he evidently considered important, but we could not understand him. We learned afterwards at the fort, that it related to the death of a chief by violence on the sea coast; this had given rise to a rumour of the death of myself, and afterwards of Dr. Richardson, which occasioned us, for a time, much anxiety. The weather, on this and several days, was remarkably fine; berries of various kinds were very abundant on the banks, and quite ripe. By the aid of the tracking line, with the occasional use of the oars and sails, we proceeded up the river at a quick rate, and reached Fort Good Hope, at half-past four on the 7th. In consequence of the above-mentioned rumour, I requested Mr. Bell, the gentleman in charge of the fort, to despatch two of the Loucheux as quickly as possible to the eastern mouth of the river, in order to gain any information the Esquimaux could give regarding Dr. Richardson's party; and, that the messengers might not be delayed by hunting on the way, I left a bag of pemmican for their use. We were sorry to learn that there was some apprehension of a serious quarrel arising between the upper and lower Loucheux, in consequence of one of Barbue's sons having killed his wife, a woman of the latter tribe.

Tuesday, 5th.

Thursday, 7th.

We quitted Fort Good Hope at noon on the 8th, arrived at the entrance of Bear Lake River on the 16th, and on the 21st reached Fort Franklin, where we had the happiness of meeting our friends in safety. The eastern detachment had arrived on the 1st of September, after a most successful voyage; and Dr. Richardson being anxious to extend his geological re-

Friday, 8th.

Thursday, 21st.

searches, as far as the season would permit, had gone in a canoe to the Great Slave Lake, having previously sent a report of his proceedings, to meet me at Fort Good Hope, in case of our being obliged to return by the Mackenzie; but the bearer of them passed us without being seen. Having read Mr. Kendall's journal, I drew up a brief account of the proceedings of both parties for the information of His Majesty's Government, and transmitted it by canoe, to Slave Lake on the following morning.

The distance travelled in the three months of our absence from Fort Franklin, amounted to two thousand and forty-eight statute miles, of which six hundred and ten were through parts not previously discovered.

I cannot close this account of our sea voyage without expressing the deep obligation I feel to Lieutenant Back for his cordial co-operation, and for his zealous and unwearied assiduity during its progress. Beside the daily delineation of the coast in the field book, the service is indebted to him for numerous drawings of scenery, as well as of the natives; and for an interesting collection of plants. My warmest thanks are likewise due to the men of my party, who met every obstacle with an ardent desire to surmount it, and cheerfully exerted themselves to the utmost of their power. Their cool, steady conduct is the more commendable, as the sea navigation was entirely novel to the whole, except the seamen Duncan and Spinks, and Hallom, the corporal of Marines. The Canadian voyagers, Felix and Vivier, first saw the ocean on this occasion.

The following Chapters contain the narrative of the proceedings of Dr. Richardson in his own words; and I embrace this opportunity of conveying my sincere thanks to him, to Mr. Kendall, and to their respective crews. I may be allowed to bear my testimony to the union of caution, talent, and enterprise in the former, which enabled him to conduct, with singular success, an arduous service of a kind so foreign from his profession and ordinary pursuits; and to the science and skill, combined with activity, of Mr. (now Lieutenant) Kendall, which must heighten the character he has already obtained for general ability and energy in his profession. I must not omit to state, that these officers describe the conduct of their crews to have been excellent.

ABSTRACT of the Mean Temperature for each Day during the Voyage along the Sea Coast west of the Mackenzie, and on the return to Bear Lake.

1826. Date.	Daily Mean	Wind and Weather.	Situation.
July.	°		
1	52.8	NNW, WNW, moderate, gloomy	Fort Good Hope.
2	58.3	West, fresh, clear	Mackenzie River, betwn. lat. 67° 28′, & 60° 53′ N., longitude 130½° & 136½° W.
3	50.3	WNW, fresh, clear	
4	55.8	West, SSW, N, light, gloomy	
5	53.7	SW, NE. moderate, gloomy, foggy	
6	45.1	NNW, ENE. fresh, moderate, rain	
7	41.6	SE, moderate, clear	
8	Not regist'd. Thermom. stolen by Esquimaux.		Mouth of the Mackenzie.
9		ENE, strong, fog and rain	Between the Mackenzie & Herschel Isld. lat. 68° 53′ & 60° 34′ N., long. 136° 19′ & 139° 5′ W.
10			
11			
12	51.6	EbyN. fresh, gloomy	
13	53.3	Variable, fog and rain	
14	50.5	Calm, rain, ENE, moderate, clear	
15	48.6	Calm, clear, NW, moderate, foggy	
16	47.3	SSE, moderate, snow, fog	
17	44.8	NW, North, moderate, hazy	Herschel Island.
18	43.6	NW, moderate, clear	Between Icy Reef and Herschel Island, latitude 69° 34′ & 69° 44° N., longitude 139° 5′ and 141° 30′ W.
19	43.4	NW, moderate, heavy rain and fog	
20	39.3	NW, fresh, fog	
21	51.3	East, SE, clear	
22	58.5	SE, light, clear	
23	51.6	West, calm, East, clear	
24	45.6	Calm, variable, clear	
25	42.0	West, light. calm, foggy	
26	44.3	Calm, NW, fog	
27	41.4	West, NW, moderate, fog	
28	43.2	ENE, light, gloomy	
29	41.6	ENE, strong, misty	
30	40.3	ENE, fresh, moderate, clear	
31	42.7	NE, moderate, clear, fresh and foggy	
Mean	47.61		
Aug.			
1	42.0	NE, gale, foggy	Between Icy Reef & Flaxman's Island, lat. 69° 44′ & 70° 11′ N., long. 141° 30′ & 145° 50′ W.
2	44.6	ENE, strong, moderate, clear	
3	44.1	ENE, moderate, clear	
4	40.7	East, moderate, clear	
5	42.6	Calm, WbyN, moderate	
6	43.2	Calm, ESE, light, clear	
7	42.8	ENE, fresh, clear	
8	42.9	ENE, strong, fog	Foggy Island, lat. 70° 16′ N. longitude 147° 38′ W.
9	41.6	NE, strong, fog	
10	39.5	ENE, strong, fog	
11	41.1	NE, moderate, fog	
12	41.1	East, moderate, very foggy	
13	41.6	NE, strong, foggy	
14	41.3	ENE, NE, moderate, foggy	
15	38.1	NE, fresh, hazy	

ABSTRACT of the Mean Temperature for each Day during the Voyage along the Sea Coast west of the Mackenzie, and on the return to Bear Lake.

1826. Date.	Daily Mean	Wind and Weather.	Situation.
Aug.	°		
16	35 .0	ENE, fresh, foggy	Return Reef, lat. 70° 26′ N. lg. 148° 52′ W.
17	37 .4	NE, gale, very foggy	
18	36 .2	NE, strong, clear	
19	36 .4	NE, strong, foggy	Foggy Island.
20	36 .4	NE, fresh, foggy	
21	35 .7	NNE, North, moderate, clear	Between Foggy Island & the Mouth of the Mackenzie, lat. 70° 16′ and 68° 53′ N. lon. 147° 38′ and 136° 19′ W.
22	37 .6	North, NE, light, clear	
23	41 .0	Calm, clear	
24	39 .4	Calm, clear, foggy in the night	
25	41 .2	Calm, fog, NE, light, ESE, strong	
26	39 .6	WNW, NW, heavy gale, snow, sleet,	
27	39 .8	Calm, ESE, light, clear	
28	43 .0	SW, strong, clear	
29	52 .5	SSW, heavy gale	
30	45 .6	NW, Heavy gale, rain	Mackenzie River.
31	42 .4	Calm, SE, gloomy	
Mean	40 .85		
Sept.			
1	38 .3	NW, gale, snow	Mackenzie River.
2	38 .6	NW, strong, clear	
3	41 .1	Calm, moderate, SE, clear	
4	41 .3	SE, NW, moderate, clear	
5	45 .9	SE, light, clear	
6	51 .0	Variable, light, clear	
7	44 .8	SE, light, NW. strong	
8	41 .0	NW, strong, snow	
9	39 .3	East, moderate, clear	
10	45 .8	SE, light, clear	
11	45 .8	NW. moderate, rain	
12	37 .3	NW, moderate, gloomy	
13	37 .2	Calm, SE, light, clear	
14	37 .9	ESE, moderate, clear	
15	42 .7	Calm, moderate, fresh, gloomy	
16	44 .5	Variable, light, gloomy	
17	36 .9	Variable, moderate, rain	
18	29 .4	NW, fresh, gloomy	
19	24 .6	NW, moderate, gloomy	
20	29 .2	ESE, fresh, clear	
21	31 .1	ENE, fresh, clear	Fort Franklin.
Mean	39 .22		

NOTE.—The thermometer used in this register, was compared with those in use at Fort Franklin during ten days after our return, and found to coincide with them.

DR. RICHARDSON'S NARRATIVE

OF THE PROCEEDINGS OF THE EASTERN DETACHMENT OF THE EXPEDITION.

CHAPTER I.

Leave Point Separation and descend the Eastern Channel of the Mackenzie —Arrive at Sacred Island—Esquimaux Graves—Interview with the Natives; their thievish disposition—Attempt to gain possession of the Union —Heavy Gale—Find Shelter in Refuge Cove—Low Coast—Mirage— Stopped by Ice at Point Toker—Reach the Sea.

July 4th. THE two parties of which the Expedition was composed, having spent the evening of the 3rd of July in cheerful conversation about their future prospects, prepared to separate on the morning of the 4th. By six o'clock all the boats were stowed; and Captain Franklin, Lieutenant Back, and their party, had committed themselves to the stream in the Lion and Reliance; while the Eastern Detachment, drawn up on the beach, cheered them on their departure with three hearty huzzas. The voices of our friends were heard in reply until the current had carried their boats round a projecting point of land, when we also embarked to proceed on our voyage. Our detachment was composed of twelve individuals, distributed in two boats, named the Dolphin and Union.

IN THE DOLPHIN.	IN THE UNION.
Dr. Richardson.	Mr. Kendall.
Thomas Gillet, *Coxswain.*	John M'Leay, *Coxswain.*
John M'Lellan, *Bowman.*	George Munroe, *Bowman.*
Shadrach Tysoe, *Marine.*	William Money, *Marine.*
Thomas Fuller, *Carpenter.*	John M'Duffey.
Ooligbuck, *Esquimaux.*	George Harkness.

The instructions we received were, to trace the coast between the Mackenzie and Coppermine Rivers, and to return from the latter overland to Great Bear Lake. Ice was the only impediment we dreaded as likely to prove an obstacle to the execution of these orders. We knew that the direct distance between the two rivers did not amount to five hundred miles; and, having provisions for upwards of eighty days stowed in the boats, we were determined not to abandon the enterprize on light grounds, especially after we had seen the friends that had just parted from us embark with so much cheerfulness in their more arduous undertaking.

On leaving Point Separation we pulled, for two hours, against the current, to regain the entrance of the "Middle Channel," which was first explored by Mackenzie, on his way to the sea, in 1798, and more perfectly surveyed by Captain Franklin, on his voyage to Garry's Island, last autumn. It has a breadth of nearly a mile, and a depth of from three to five fathoms; though in one place, where there was a ripple, the sounding lead struck against a flat bed of stone in nine feet water. Having proceeded about ten miles in this channel, we entered a branch flowing to the eastward, with the view of tracing the course of the main land. Mackenzie, on his return from the sea by this route, observed many trees having their upper branches lopped off by the Esquimaux, and we saw several such trees in the course of the day. The lands are low and marshy, and inclose small lakes which are skirted by willows. The summits of the banks are loaded with drift-timber, showing that they are all inundated by the spring floods, except a few sandy ridges which bound the principal channels, and which are clothed with well-grown white spruce trees, Our voyage amongst these uninteresting flats was greatly enlivened by the busy flight and cheerful twittering of the sand-martins, which had scooped out thousands of nests in the banks of the river, and we witnessed with pleasure their activity in thinning the ranks of our most tormenting foes the musquitoes. When our precursor, Sir Alexander Mackenzie, passed through these channels on the 10th of July, 1789, they were bounded by walls of ice veined with black earth, but the present season was so much milder, that the surface of the banks was every where thawed.

An hour before noon we put ashore to cook our breakfast, near a clump of spruce trees, where several fires had recently been made by a party which had left many foot-prints on the sand; probably a horde of Esquimaux, on their return from trading with the Indians at the Narrows. A thunder storm that obscured the sky, prevented Mr. Kendall from ascertaining the latitude at noon, which was the hour we chose for breakfast throughout the voyage, in order to economize time, as it was necessary to land to obtain the meridian observation of the sun. In the afternoon we continued to descend the same channel, which has a smooth and moderately rapid current, and a general depth of two or three fathoms. At four P.M. we obtained a view of a ridge of land to the eastward, which we have since learned is named by the natives the Rein-Deer Hills, and at seven encamped near two conical hills of limestone, about two hundred feet high, and clothed with trees to their tops. The length of the day's voyage was forty-two miles. We selected a sandy bank, covered with willows sixteen feet high, for our encamping place; and here again we found that a party of Esquimaux had lately occupied the same spot, the ashes of their fires being still fresh, and the leaves of the willow poles to which they had attached their nets, unwithered. Before we retired to bed, the arms were examined, and a watch was set; a practice which we kept up for the remainder of the voyage. Much rain fell in the night.

Wednesday, 5th. On the 5th we embarked at four in the morning, and soon afterwards, the channel conducting us to the base of the Rein-Deer Hills, Mr. Kendall and I ascended an eminence, which was about four hundred feet high. Its summit was thinly coated with gravel, and its sides were formed of sand and clay, inclosing some beds of brownish-red sandstone, and of gray-coloured slate-clay. Clumps of trees grew about half way up, but the top produced only a thin wiry grass. At eleven A.M. we landed to breakfast, and remained on shore until noon, in the hope of obtaining an observation for latitude, but the sun was obscured by clouds. In the afternoon I had an extensive view from the summit of a hill of flat alluvial lands, divided into islands by inosculations of the channels of the river, and bounded, at the distance of about forty miles to the westward, by the Rocky Mountains. As we advanced to the northward, we perceived the trees to diminish in size, becoming more scattered, and ascend a shorter way up the sides of the hills, and they altogether terminated in latitude 68° 40′, in an even line running across the islands; though one solitary spruce fir was seen in 68° 53′. Perhaps the lands to the northward of this abrupt line were too low and wet for the growth of the white spruce, the tree which attains the highest latitude on this continent.

We pitched our tents for the night on the site of another Esquimaux encampment, where a small bit of moose deer's meat was still attached to a piece of wood at the fire-place; and we saw, from the tracks of the people and dogs in the sand, that a party had left the river here to cross the Rein-Deer Hills. From information obtained through the Sharp-eyed, or Quarreller tribe of Indians, this appears to be one of the Esquimaux routes to a large piece of brackish water named Esquimaux Lake, and alluded to by Mackenzie in several parts of his narrative. The length of our voyage this day was forty-four miles, and our encampment was opposite to an island named by Captain Franklin after William Williams, Esq., late governor of Prince Rupert's land. We observed here an unusually large spruce tree, considering the high latitude in which it grew; it measured seven feet in circumference, at the height of four feet from the ground. A hole was dug at the foot of the hill, in sandy soil, to the depth of three feet without reaching frozen ground.

Thursday, 6th.

On the 6th, heavy and continued rain delayed our embarkation until ten o'clock in the forenoon, and the weather, during the rest of the day, was hazy, with occasional showers of small rain. Before leaving the encampment, we lopped the branches from a tree, and suspended to it a small kettle, a hatchet, an ice-chisel, and a few strings of beads, together with a letter written in hieroglyphics, by Mr. Kendall, denoting that a party of white people presented these articles to the Esquimaux as a token of friendship.* As we advanced, we came to the union of several ramifications of the middle channel with the eastern branch of the river, and the breadth of the latter increased to two miles; its depth of water being rarely less than three fathoms. In latitude 69°, the eastern channel of the Mackenzie makes a turn round the end of the Rein-deer-hills which terminate there, having previously diminished in height to about two hundred feet. At the commencement of this turn, there is a small island nearly equal to the main land in height, and appearing when viewed from the southward, to be a continuation of it. Its position pointing it out to be the one described by Mackenzie as possessing "a sacred character," and being still a burial place of the Esquimaux, I named it Sacred Island. We saw here two recent,

* As the reader may desire to know what hieroglyphics were used to express our intentions, a copy of the letter is annexed.

and several more ancient graves. The bodies were wrapped in skins closely covered with drift-wood, and laid with their heads to the west; so that the rule mentioned by Captain Lyon in his account of Melville peninsula, does not obtain on this part of the coast; for there none but the bodies of infants are placed in that direction. Various articles, such as canoes, sledges, and fishing nets, were deposited near the graves.

Sacred Island is formed entirely of layers of fine sand of different colours, covered by a little vegetable mould. One of its sides being steeply escarped by the waves, showed its structure completely. Amongst the vegetable productions of this spot, we noticed the perennial lupine, the narrow-leaved epilobium, and some currant bushes in full flower, and growing with great luxuriance. From its summit we had a view of the river flowing in many channels, both to the eastward and westward. The islands lying in sight to the westward are low, and apparently inundated when the river is flooded; but to the eastward, there are many islands having hummocks as high as Sacred Island, and judging from those that were near, they are, like it, composed of sand. The channels surrounding the island appear to be shallow.

After leaving that island, we steered along the main shore to a sandy point about four miles distant, and encamped near a very recent resting place of a large party of Esquimaux, not fewer than ten fires having been made since the heavy rain of the morning. There was also vestiges of five or six winter-houses on this point. Richard's Island, which was named in honour of the Governor of the Bank of England, forms the opposite bank of the channel here, and exhibits, like the neighbouring islands, some sandy hummocks and cliffs. The length of the day's voyage was twenty-five miles, and our encampment was situated in lat. 69° 4′ N., long. 134° 10′ W.

Friday, 7th. We embarked on the morning of the 7th at four o'clock, in cold, hazy weather, and soon came to a point of Richards' island, on which there were four or five Esquimaux tents, with several skin canoes, and boats lying on the beach. I had previously arranged that on our first interview with the Esquimaux, I was to land with Ooligbuck, whilst Mr. Kendall kept the boats afloat ready to lend us such aid as might be required; conceiving that this was the best way of inspiring the natives with confidence, should they be distrustful, or of securing freedom of action to our crews should they prove unfriendly. The muskets were kept in the arm-chest out of sight, but ready for instant use. As we drew near the point, two women, who were walking along the shore, looked at us with amazement for some minutes, and then ran into the tents and alarmed their inmates. Several men instantly rushed out, nearly naked, with their bows and quivers in their hands, making furious gestures and apparently much frightened. I desired Ooligbuck to speak to them, and called to them myself in their own language that we were friends; but their terror and confusion was so great, that they did not appear to comprehend us. I then took a few beads, files, and knives, in my hand, and landing with Ooligbuck, made some presents to the men, and told them I was come to trade. The moment I mentioned the word "trade" (*noowœrlook*), their fears subsided, and they sent away their bows, but retained their long knives; those that were clothed thrusting them into their pockets or up their sleeves. An old woman who seemed to have greater self-possession than the rest, and to understand my meaning more readily, ran and fetched some dried fish, for which I gave her beads; and the others then began to manifest an eager desire of exchanging their fish for any thing that I offered. More people coming from the tents, a crowd was formed, who obtained all the trading articles I had brought on shore. As their surprise subsided, their boldness and clamour increased, and some few of them began again to use threatening expressions and gestures, either from a dislike to strangers coming into their country, or for the purpose of intimidation and extortion, When the interview assumed this disagreeable character, Ooligbuck said that they were very bad people, and entreating me to embark, took me on his back and carried me on board. At the same time, several of the natives ran into the water and attempted to drag the boat ashore, but on my calling to them they desisted. One fellow, whose countenance, naturally disagreeable, had been rendered hideous by the insertion of a large brass thimble into a perforation in the under lip, seized upon our tea-kettle, and endeavoured to conceal it under water, but being seen from the Union, he was made to return it.

When we left the shore, all the males, twenty-one in number, embarked in their small canoes or kaiyaks and accompanied us; and in less than a quarter of an hour, the women had struck the tents and embarked them, together with their children, dogs, and luggage, in their row boats or oomiaks, and were in close pursuit. For a time we proceeded down the river together in an amicable manner, bartering beads, fire-steels, flints, files, knives, hatchets, and kettles, for fish, adzes, spears, and arrows. The natives seemed to have a correct idea of property, and showed much tact in their commerce with us; circumstances which have been held by an eminent historian to be evidences of a considerable progress towards civilization*. They were particularly cautious not to glut the market by too great a display of their stock in trade; producing only one article at a time, and not attempting to out-bid each other; nor did I ever observe them endeavour to deprive one another of any thing obtained in barter or as a present. As is usual with other tribes of Esquimaux, they asked our names and told us theirs, a practice diametrically opposite to that of the Indians, who conceive it to be improper to mention a man's name in his presence, and will not, on any account, designate their near relatives, except by some indirect phrase. They showed much more curiosity respecting the construction of our boats than any of the tribes of Indians we had seen, and expressed great admiration of the rudder, soon comprehending its mode of action, although it is a contrivance of which they were previously ignorant. They were incessant in their inquiries as to the use of every thing they saw in our possession, but were sometimes content with an answer too brief to afford much explanation; as in the following instance. Ooligbuck had lighted his pipe and was puffing the smoke from his mouth, when they shouted "*ookah, ookah,*" (fire, fire,) and demanded to be told what he was doing. He replied with the greatest gravity, "*poo-yoo-al-letchee-rawmah*" (I smoke); and this answer sufficed. On my referring to an Esquimaux vocabulary, Ooligbuck, in answer to their questions, told them that the book spoke to me, when they entreated me to put it away. I afterwards detected the rogue with the brass thimble endeavouring to steal this book, and placed it, as I thought, out of his reach; it was missing in the evening, but I never ascertained whether it had been purloined by the Esquimaux or had fallen overboard in moving some of the stores. Seeing me use my pocket telescope, they speedily comprehended its use, and called it "*eetee-yawgah,*" (far eyes) the name that they give to the wooden shade which is used to protect their eyes from the glare of the snow; and which, from the smallness of its aperture, enables them to see distant objects more clearly. Of our trading articles, light copper kettles were in the greatest request, and we were often asked for the long knives which are used for flinching whales. It is creditable to the Esquimaux habits of cleanliness, that combs were in great demand, and we saw wooden ones of their own manufacture, not dissimilar to ours in form. I distributed looking-glasses to some of the young men, but they were mostly returned again, although I do not know on what account.

* Robertson's *History of America.*

This party of Esquimaux, being similar in features and dress to the tribe seen by Captain Franklin, and not differing materially from the Esquimaux inhabiting Melville peninsula which have been so fully described by Captains Parry and Lyon, it is not necessary to enter into any detail here on those points. Ooligbuck's dialect and theirs differed a little, but they mutually understood one another. I observed that they invariably sounded the letter *m* instead of *g*, when in the middle of a word, calling Ooligbuck, Oolimbauk. Ooligbuck's attempts to pronounce "Doctor" were sufficiently imperfect, but to our visitors, the word seemed utterly unattainable, and they could designate me only by the term *Eheumattak* or chief. They succeeded better with the names of some of the men, readily naming Tysoe, and calling Gillet "*Hillet.*" The females, as they passed in their oomiaks, bestowed on us some glances that could scarcely be misconstrued,—their manners, in this respect, differed widely from those of the Indian women, who have a modest and even shy demeanour. Some of the young girls had a considerable share of beauty, and seemed to have spared no pains in ornamenting their persons. Their hair was turned up in a neat knot, on the crown of the head, and a lock or queue, tied by a fillet of beads, hung down by the ears, on each side. Mr. Nuttall, in his account of the Quapaws or Arkansas, mentions that the unmarried women wear their hair braided into two parts, brought round to either ear in a cylindrical form and ornamented with beads; and a similar attention to head-dress is paid by some of the Indian women inhabiting the borders of the great Canada lakes, and also by the Tawcullies or Carriers of New Caledonia;* but the females of all the tribes of Indians that we saw in our route through the northern parts of the fur countries, suffer their hair to hang loose about their ears, and, in general, adorn their persons less than the men of the same tribes. The Esquimaux women dressing better, and being required to labour less, than the Indian females, may be considered as a proof that the former nation has made the greater progress towards civilization; and I am of opinion that the Esquimaux would adopt European habits and customs much more readily than the Indians.

Though there are many circumstances which widely distinguish the Esquimaux from their Indian neighbours, they might all, possibly, be traced to the necessity of associating in numbers for the capture of the whale, and of laying up large hoards

* *Harmon's Journal*, p. 288.

of blubber for winter consumption. Thus have they been induced to build villages for their common residence, and from thence have originated those social habits which are incompatible with the wandering and precarious life of an Indian hunter. It would lead, however, to too long a digression, were I to enter into details on this subject, and I resume, therefore, the narrative of the voyage.*

In the course of the morning we came to several other encampments, one of them consisting of nine tents; and each party no sooner learnt who we were, than they embarked bag and baggage and followed us. Some of the new comers were shy, and kept aloof, but in general they were too forward. Emboldened by their increase of numbers, they gradually became more daring, and running their kaiyacks alongside, laid hold of the boat's gunwale, and attempted to steal any thing within their reach. To lessen their opportunities of annoying us, I was obliged to keep the crews constantly rowing, for when we attempted to rest, three or four fellows would instantly seize the opportunity of lifting the blades of the oars and pushing their kaiyacks alongside, whilst others would cling on by the bows and quarters, nor could they be dislodged without much trouble. They manifested great cunning and dexterity in their pilfering attempts, and frequently acted in concert. Thus, one fellow would lay hold of the boat with both his hands; and while the coxswain and I were disengaging them, his comrade on the other side would make the best use of his time in transferring some of our property into his canoe, with all the coolness of a practised thief. The smaller things being, however, put as well out of the way as possible, and a strict look-out kept, they were, in almost every instance, detected; and they restored, with the most perfect good humour, every article they had taken, as soon as it was demanded, often laughing heartily at their own want of address. They succeeded only in purloining a bag of ball, and a powder-horn, as

* The Esquimaux method of settling disputes, which we learned from Augustus, deserves to be mentioned, not only as being very different from the sullen conduct of an affronted Indian, but from its coincidence with the practice of a people widely separated from them—the native inhabitants of Sydney, in New South Wales. Mr. Cunningham, in his entertaining work on New South Wales, says, "The common practice of fighting amongst the natives is still with the *waddie*, each alternately stooping the head to receive the other's blows, until one tumbles down, it being considered cowardly to evade a stroke." The Esquimaux use the fist instead of the waddie, in these singular duels, but there is no other difference betwixt their practice and that of the New South Wales' people. Another coincidence betwixt the Esquimaux and the inhabitants of Australasia, is the use of the throwing stick for discharging their spears.

the theft was not perceived at the time. I was unwilling to check this conduct by a display of arms, because I was desirous of gaining the natives by kindness and forbearance, the more especially, as our ignorance of the state of the ice rendered it doubtful, whether we might not be under the necessity of encamping, for some time, in their neighbourhood. Had we resented their pilfering attempts too hastily, we should have appeared the aggressors, for they expressed great good-will towards, us, readily answered such questions as we were able to put to them about the course of the river, pointed out to us the deepest channels, invited us to go ashore to cook our breakfast, and even offered to provide us with wives, if we would pass the night at their tents. For very obvious reasons we declined all their invitations; but our crews being fatigued with continual rowing, and faint from want of food, we halted at one P.M., by the side of a steep bank, and breakfasted in the boats, insisting on the Esquimaux keeping aloof whilst we were so engaged.

In the afternoon we had to search for a passage amongst islands, there being no longer water enough near the main shore to float our boats. The Esquimaux undertook to guide us, but whether through accident or design, they led us, on one occasion, into a shallow channel, where we grounded on a sandbank, over which there was a strong current setting; and we had not only much difficulty in getting afloat, but had to pull, for an hour, against the stream, to regain the passage we had left. Soon after this, one of the natives made a forcible attempt to come into the Dolphin, under the pretext of bartering two large knives which he held in his hand; and the dexterity with which he leaped from his kaiyack was remarkable. There were three other kaiyacks betwixt him and our boats, which, on his giving the signal, were, by their owners laying their broad paddles across, instantly converted into a platform, over which he ran with velocity and sprang upon the stern seat of the Dolphin, but he was immediately tumbled out again. Judging from the boldness of this fellow's behaviour, and the general tenour of the conduct of the natives, that sooner or later they might be tempted to make an attack upon us, I adopted, as a measure of precaution, the plan of purchasing their bows, which are their most powerful weapons. They were at first unwilling to part with them; but finding that we would take nothing else in exchange for the articles we had to dispose of, they ultimately let us have a good number. The Esquimaux bows are formed of spruce-fir, strengthened on the back by cords made of the sinews of the rein-deer, and would have been prized, even beyond their favourite yew, by the archers of Sherwood. They are far superior to the bows of the Indians, and are fully capable of burying "the goose-wing of a cloth-yard shaft" in the heart of a deer.

Several of the young men tried the speed of their kaiyacks against our boats, and seemed to delight in showing us how much their little vessels excelled ours in velocity. Towards evening the women's oomiacks had all gone ahead, and we were given to understand that they were about to encamp for the night. Thinking that they would choose the best route, we followed them into a channel, which proved too shallow; and when we put about to try another, the natives became more urgent than ever that we should land and encamp along with them. Just as we were about to enter a passage which the Esquimaux, doubtless, knew was deep enough, and led by the shortest route to the sea, the Union grounded upon a bank, about half a bow-shot from the shore. Seven or eight of the natives instantly jumped out of their kaiyacks, and laying hold of the boat's bow and steering-sweep, attempted to drag her ashore. They were speedily joined by others, who hurried from the beach with knives in their hands; and Mr. Kendall seeing that he would almost immediately be surrounded by a force too great to permit his men to act, called to me that he should be obliged to fire. Fully aware of the necessity of prompt measures, I answered that he was at liberty to fire if necessary. Upon which, snatching up his fowling-piece, he presented it at three of the most daring who had hold of the sweep-oar, and his crew who were now in the water endeavouring to shove the boat off, and struggling with the natives, jumped on board and seized their muskets. The crew of the Dolphin likewise displayed their arms and stood ready, but I ordered that no individual should fire until called upon by name. They were, however, the instant that a shot was fired from the Union, to lay the Dolphin aground alongside of her, that thus we might present only two assailable sides to the enemy. Happily there was no occasion to fire at all; the contests of the Esquimaux with the Indians had taught them to dread fire-arms, and on the sudden sight of every man armed with a musket, they fled to the shore. Until that moment we had kept our guns carefully concealed in the arm-chest, to prevent any of the natives from snatching them away and disarming us, and also that they might not deem our intentions to be other than pacific.

I do not believe that the natives had matured a plan of attack, but the stranding of a boat on their own shore was too great a temptation to be resisted. Some individuals had previously shown unequivocal signs of good feeling towards us, such as bringing back the Union's sweep-oar, which had slipped from the coxswain's hands; and also in pointing out the channel we afterwards pursued to the sea, as preferable to the one which the oomiacks had taken. Even the better-disposed, however, would, doubtless, have joined the others, had they began to plunder with success; for they told us in the forenoon that there was no one of their horde acknowledged as a chief. It is probable that the Esquimaux were doubtful as to the sex of some of our party, until they saw them prepare for battle. None but women row in their oomiacks, and they had asked Ooligbuck if all the white women had beards.

The crews on this occasion behaved with a coolness and resolution worthy of the utmost praise, executing without the slightest confusion the orders they received. Mr. Kendall acted with his usual judgment; and his prudence and humanity, in refraining from firing, merit the highest encomiums. The Union being speedily set afloat by her crew, we pulled together through a wide channel, three feet deep. The spot where this transaction took place has been named Point Encounter, and is in latitude 69° 16′ N., and longitude 136° 20′ W.

The Esquimaux seemed to hold a consultation on the beach after we left them; but, as none attempted to follow us immediately, we enjoyed the respite from their forwardness and clamour, which had become very harassing, particularly to Mr. Kendall and myself, who had other duties to attend to. He had full occupation in surveying and delineating the route; and as the Dolphin led the way through a shoal and intricate navigation, it was requisite that I should keep the sounding-lead constantly going, and be on the watch for any change in the appearance of the current which might indicate shoal water, the smallness of our crews preventing me from appointing any man to that service. In about an hour after leaving Point Encounter, we observed ten kaiyacks coming towards us from a cluster of islands; they soon overtook us, but kept at a reasonable distance, and no longer gave us any trouble by coming alongside. We wished to show that we had no desire to hurt them, notwithstanding their past conduct, and, therefore, began again to trade with them; yet we were naturally anxious that they should leave us before we encamped, because, from the fleetness of their kaiyacks, they could soon collect a great number of their countrymen, and give us much annoyance in the night. Our wishes were seconded by a fresh breeze of wind springing up and enabling us to set the sails, by which the crews enjoyed a rest, after fourteen hours' labour at the oars; and the Esquimaux had greater difficulty in keeping up with the increased velocity of our boats. Thinking that they would quit us as soon as they lost the hope of getting more goods, I desired Ooligbuck to tell them I would trade no more, and they accordingly, one by one, dropped behind and left us. Three followed us longer than the others, and as they were not of the party which attacked the Union, and had hitherto received nothing from us, I made each of them a small present of beads and fire-steels, when they also took leave, calling out "*teymah, peechaw-ootoo,*" "friendship is good."

We learned in the course of the day, from the natives, that they call themselves *Kitte-garræ-oot*, (inhabitants of the land near the mountains,) and that they were now on their way to a place favourable for the capture of white whales, as in the sea, which they said was many days' march distant, there was too much ice to take the black whales at this season. It also appeared that they annually ascend to the Narrows of Mackenzie River, for the purpose of trading with the Quarrellers, and were accustomed to spend their summers in a large lake of brackish water, (Esquimaux Lake,) lying to the eastward, where they occasionally meet parties of Loucheux. They informed us that the land to the eastward of Encounter Point is a collection of islands, and that there were many of their countrymen fishing in the rivers which separate them. They had heard of the Esquimaux at the mouth of the Coppermine River, and knew them by their name of *Naggæ-ook-tor-mæ-oot*, (or Deer-horns,) but said they were very far off, and that they had no intercourse with them; adding, that all the inhabitants of the coast to the eastward were bad people. They knew white people by the name of *Kabloonacht*, and Indians by that of *Eitkallig*, the same appellations that are used by the Esquimaux of Hudson's Bay; but their name for the black whale was different from that given to it by Ooligbuck; and they also gave names to some of their utensils which he had never heard before. Ooligbuck was not of much use as an interpreter, in our intercourse with these people, for he spoke no English; but his presence answered the important purpose of showing that the white people were on terms of friendship with the distant tribes of Esquimaux. As a boatman he was of the greatest service, being strongly attached to us, possessing an excellent temper, and labouring cheerfully at his oar.

We could not ascertain the numbers of Esquimaux we saw in the course of the day, because they were always coming and going, but we passed at least thirty tents, and had reason to believe that on some of the islands there were tents which we did not see. Four grown people is, perhaps, the average number of the inhabitants of each tent. A short time before the attack on the Union, I counted forty kaiyacks round the two boats.

The wind freshened, and the night began to look stormy, as we stood across a wide sound which was open from the N.W. to the N.E., and had a depth of water varying from three to seven feet. White whales were seen; and some of the crew thought the water tasted brackish. About nine P. M. a drizzling rain came on, attended with very dark weather, which induced us to make for a round islet, with a view of encamping, and securing the boats for the night; it was skirted by shoals that prevented us from landing, and we therefore anchored the boats by poles stuck in the mud, raised the coverings of the cargo on masts and oars, so as to turn off the rain; and after eating our supper and setting a watch, we endeavoured to get some repose by lying down in our clothes, wet as they were. We had scarcely laid down, however, before the wind changed and began to blow with violence directly on the shore, so as to render it necessary for us to shift our situation without delay. An attempt was made to row the boats round to the other side of the islet, but they drifted upon the shoals in spite of the exertions of the crew, and began to strike violently. In this perilous situation we perceived some smooth water to leeward, upon which setting the foresails, the boats were pushed over a sandy bar into two fathoms water. We then stood towards the eastern shore, and keeping in deep water, entered a small inlet, which received the name of Refuge Cove; where having made fast the boats to the beach, pitched a tent on the shore, and set a watch, we attempted a second time to obtain some rest.

Saturday, 8th.

We were not, however, destined to enjoy much repose that night, for we had scarcely overcome the chilliness occasioned by lying down in wet clothes, when the Union broke from her moorings in a violent gust of wind, and began to drive across the inlet towards the lee-shore, on which there was a considerable surf. Mr. Kendall and one of the crew, who were sleeping on board, to be ready in case of accident, lowered the covering with the utmost expedition, and taking the oars, kept her from driving far, until the rest of the party arrived to their assistance in the Dolphin. The boats were brought to the beach and secured, and we had again retired to rest, when the tent-pegs, although loaded with drift timber, were drawn up by the force of the wind, and the tent, drenched with rain, fell upon us. It was in vain to attempt to sleep after this, benumbed as we were by the coldness of the weather; but the rain ceasing about four in the morning of the 8th, we were enabled to make a good fire, and dry our clothes. The cargo of the boats was then landed, the wet packages spread out to dry, and the boats were drawn upon the beach so as to form, with the baggage, a three-sided breast-work, to which we could retreat, should the Esquimaux pay us a hostile visit. These arrangements being made, the tent was removed to a more sheltered spot, and we slept quietly until ten o'clock in the morning. In the night an accident happened to Mr. Kendall, which might have had fatal consequences, and alarmed us at the time exceedingly. The point of a small two-edged knife which he wore in a sheath slung from his neck, was, by his falling against one of the tent-poles, forced through the sheath into his side, exactly in the region of the heart. Through the mercy of Providence, its progress was arrested by one of the ribs, and the wound healed in the course of a few days. At noon a meridian observation was obtained, which placed the mouth of Refuge Cove in latitude 69° 29′ N.; and the sun's bearing showed the variation of the magnetic needle to be 49½° easterly. The length of our voyage the preceding day was fifty-seven miles. Refuge Cove has an irregular form; its length is about two miles and a half, and its greatest width one mile. It is upwards of two fathoms deep at the entrance, and for some distance within; but a bar runs from Shoal Islet to its north side. Its shores are flat and sandy, but here and there hummocks rise abruptly to the height of one hundred feet, resembling the downs on the Norfolk coast. The sandy hummocks are bound together by the creeping fibrous roots of a species of grass, named *Elymus mollis;* and many of them are covered by a coat of black vegetable mould. Ruins of Esquimaux houses, that appeared to have been deserted for many years, were scattered along the borders of the cove, and much drift-timber lay on the low grounds. We saw some ducks and geese, and two of the crew went to hunt round the harbour for deer, but they had no success.

The wind having moderated in the evening, we prepared to resume the voyage, and had begun to load the boats, when I thought I saw a *kaiyack* paddle across the mouth of the cove. It was followed by many others, that were in succession lost behind the point, with the exception of one which seemed to return and look into the inlet. I concluded that the natives were in search of us; and, as it was desirable to have all the cargo on board when they arrived, the utmost despatch was used in loading the boats. Before this operation was completed, Mr. Kendall, on attentively examining one of the objects with his telescope, suggested that it was not a kaiyack; and accompanying me to a sandy eminence nearer the entrance of the cove, we ascertained that the whole was an optical deception, caused by the haze of an easterly wind magnifying the stumps of drift wood, over which the surf was rolling. The imagination, no doubt, assisted in completing the resemblance, but the deception, for a few minutes, was perfect.

We quitted Refuge Cove at nine o'clock in the evening, and rounding Shoal Islet, steered to the northward along the coast. The circuit of Shoal Islet was made because there was too little water to float our boats over the bar, which we had crossed the preceding evening. The temperature of the air on leaving the cove, was 36°, but it fell at midnight to 32°; and the night proved fine. When resting on our oars, the boats were drifted to the westward, by a current which we ascertained, by subsequent observations, to be the flood tide.

Sunday, 9th.

After pulling along the coast for some time, the ice-blink appeared in the horizon, and about one o'clock in the morning on the 9th, we could perceive a stream of ice lying at the distance of eight or nine miles from the shore, and inclosing several small ice-bergs. At four o'clock, a northerly breeze springing up, brought a quantity of loose ice down upon us, and we made for the shore. This part of the coast is skirted to the distance of two miles by flat sands, on which there is not more than a foot or eighteen inches of water. The depth of water gradually increases to four fathoms, which it attains at the distance of six or seven miles from the shore, and the heavy ice we saw outside, showed that the depth there was considerable. Esquimaux winter-huts occur frequently on the coast, and the rows of drift-trees planted in the sand with the roots uppermost, in their vicinity, assume very curious forms, when seen through a hazy atmosphere. They frequently resembled a crowd of people, and sometimes we fancied they were not unlike the spires of a town just appearing above the horizon. We learnt by experience that the shore was more approachable at the points on which the Esquimaux had built, and we effected a landing at one of those places, when, having discharged the cargoes, we hauled the boats up, and pitched the tents. The water at our landing-place was fresh, but too hard to make tea; and at four or five miles from the shore, it was disagreeable to drink. Out of respect to Captain Toker of the Royal Navy, under whom I had once the honour to serve, his name was given to this Point.

Mr. Kendall ascertained its latitude to be 69° 38′ N.; its longitude by reckoning, 132° 18′ W.; and the variation of the magnetic needle 50½ degrees easterly. The distance rowed from Refuge Cove was about twelve miles. A tide pole was erected, by which it appeared that the ebb ran from four o'clock, the time at which we landed, until ten in the morning, producing a fall of eighteen inches; but the afternoon tide did not rise so high, and at 10h. 50′ P.M. it was low water again, the wind blowing fresh from the northward all the time.

The vicinity of Point Toker, like the rest of the lands to the eastward of Point Encounter, consists of level sands, inclosing pieces of water which communicate with the estuary of the river, and interspersed with detached conical hills rising from one to two hundred feet above the general level. These hills are sometimes escarped by the action of the water, and are then seen to consist of sand of various colours, in which very large logs of drift-timber are imbedded. They are covered by a coat of black vegetable earth, from six inches to a foot in thickness, which shows that they cannot be of very recent formation, though at some distant period they may have been formed by the drifting of moveable sands. At present, the highest floods reach only to the foot of the hills, where they deposit a thick layer of drift-timber. One straight log of spruce fir, thirty feet long, was seven feet in circumference at the small end, and twelve a short distance above the root. The branches and bark are almost always rubbed off from the drift-timber which reaches the sea, but a few of the main divisions of the root are generally left. Various instruments tied up in bundles were suspended to poles near some of the Esquimaux houses, such as spear-heads and ice chisels made from the tooth of the narwhal, and spoons of musk-ox horn. The marine animals that frequent this part of the coast, according to the information we obtained from the Esquimaux, are, the white whale, the narwhal, large and small seals, (*oggæ-ook* and *natchæ-ook*,) and a species of black whale, named *aggee-wærk*. There are also many sea-fish, of which the capeline (*ang-maggæ-ook*,) that abound on the shoals at this season, are most easily caught. The natives are unacquainted with sea-horses. Swans, Canada and white geese, and Arctic ducks, are numerous, and we killed several. Ooligbück likewise killed a rein-deer, which afforded us an agreeable change of diet.

In the evening, having assembled in one of the tents, prayers were read, a practice to which we adhered on every Sunday evening during the voyage. At 10h. 45m. P.M., I lighted a piece of touchwood with a convex lens, an inch in diameter, the altitude of the sun being then 3° 6′. It is seldom that the sun in warmer climates affords so much heat at so low an altitude.

The ice opening a little, we resumed the voyage at five o'clock in the morning of the 10th, but had not rowed above five miles, when our further progress was impeded by a ridge of grounded-ice, extending apparently far out to sea. We landed to obtain a view from a height, and took advantage of the opportunity to prepare breakfast. Whilst thus engaged, we discovered, on the opposite side of a bay which we had just crossed, two of the natives couched upon the sand, and evidently watching us; but before we had concluded our meal, they went off. On re-embarking, we went round the ice which was aground on extensive sandy spits, and then pulled in for the shore; but a fresh breeze of wind created such a swell, that we did not advance above three miles in two hours. Deeming it unadvisable to fatigue the crews, while the progress was so small, we pulled into a sandy bay, and made the boats fast to one of many large pieces of ice which were stranded on the beach, having gained since setting out in the morning, eight miles.

Monday, 10th.

Just as we made for the shore, we observed three Esquimaux regarding us from an eminence, and two others soon afterwards joined them: the latter being, as we discovered from the direction of their path over the sands, the two we had seen at breakfast-time. They retired as we drew near the beach, and on reconnoitring the neighbourhood, we discovered three skin-tents, whose owners were running off with their effects in great alarm. As we had experienced how troublesome the natives were, when relieved from their fears, we did not seek an interview at this time; and to guard against accidents from parties of them way-laying our men, I determined that, while we remained in this anchorage, the crews should land only to cook their provisions and then be accompanied either by Mr. Kendall or myself. The water at our anchorage was decidedly brackish, the beach was strewed with *sertulariæ* and other marine productions, and several white whales were seen in the offing; all which circumstances being considered as decided evidences of our having reached the mouth of the river, that event was celebrated by issuing to each of the men a glass of grog, which had been reserved for the occasion.

CONTINUATION OF THE PROCEEDINGS OF THE EASTERN DETACHMENT.

CHAPTER II.

Detention by wind—Visited by Esquimaux—Cross a large Stream of fresh Water—Winter Houses on Atkinson Island—Gale of Wind, and Fog—Run into Browell Cove—Double Cape Dalhousie—Liverpool Bay and Esquimaux Lake—Icy Cliffs—Meet another Party of Esquimaux—Cape Bathurst.

Tuesday, 11th.

THE wind blew so strongly during the 11th, that we remained in our mooring-place, landing occasionally to take a little exercise on the beach; and as it continued to freshen from the north-east in the evening, most of the ice in the offing had drifted out of sight, while a great reduction took place at the same time in the number and size of the pieces of stranded ice. One of them which had grounded about a mile outside of us, and rose fifteen feet above the water, fell over and floated away with the ebb tide. Mr. Kendall obtained a meridian observation for latitude, and afterwards took several sets of lunar distances, whose results placed our anchorage in latitude 69° 42½′ N., and longitude 131° 58′ W. In the afternoon two Esquimaux were seen walking fast over a hill, and often stopping and looking anxiously around them. About midnight two black foxes carried off the scraps of meat that had been left at our cooking-place, and buried them carefully in the sand above high-water mark. We observed that they dug separate hiding-places for each piece, and that they were careful to carry the largest bits farthest from the sea. The time spent inactively at the anchorage was so irksome, that even the movements of these animals were a subject of much interest to us, and we felt great regret when they were scared away by the talking of the men in the boats.

Wednesday 12th.

No material alteration took place in the weather on the 12th. The temperature was 45°; but from the force of the wind, and our confinement in the boats, we felt cold. In the evening two elderly Esquimaux came to us in their kaiyacks, shouting as they approached the boats, and paddling boldly alongside. They told us that they were the same two whom we had seen in the morning of the 10th watching us while at breakfast, though they had first discovered us on the 9th, and had seen Ooligbuck kill the deer, which had alarmed them greatly; they had since been to inquire about us from the party at Point Encounter, and having learnt that we were well-disposed, they had come to open a communication. In allusion, I suppose, to the attempt on the Union, they often said that the Esquimaux at the river's mouth were bad people, but that they themselves were good-hearted men; and they struck their breasts forcibly with their hands, to give energy to their assurances. They told us that a large party of their countrymen, who were at present fishing at the mouth of a river to the eastward, would soon move in this direction to kill white whales. Eetkoo-yak, the principal spokesman, invited us to go to his tents, where he said, the women would be glad to receive us; and added, that next day he would bring four of his countrymen to visit us. We made them a handsome present of iron-work; and having paid, with beads, for some dried fish, sent them away highly contented.

Thursday, 13th.

At seven o'clock in the morning of the 13th, nine Esquimaux came to us, amongst whom were our two acquaintances of yesterday. Some of the young men inquired when we were going away, and seemed to be anxious that we should depart; but our friend Eetkoo-yak gave us a pressing invitation to his tents, and wished to embark in the boats to conduct us thither. We declined his proposal, and the wind having moderated, we unmoored the boats, and rowed along the coast. The natives followed us, and soon afterwards four women and two boys came off in an oomiak, and exchanged some boots, pieces of leather, deer's meat, and fish for beads. The point on which their tents were pitched was named Point Warren after my friend Captain Samuel Warren, R. N. As we continued our course the oomiak returned to the shore, and the men also left us soon afterwards, apparently pleased with our departure; for the knowledge of the effect of our muskets seemed to have impressed them with some dread. They were tattooed across the cheeks. The tribes to the westward of the Mackenzie are described by Captain Franklin, (p. 111,) as following a different fashion in the application of this ornament.

We coasted this day a flat shore, with dry sands running off to the distance of two or three miles, and we passed within several shoals, on which some heavy ice had grounded. Only a few small streams of ice were seen, although the ice-blink was visible the whole day. Soon after rounding Point Warren, we crossed the mouth of a large river, the water being muddy and fresh for a breadth of three miles, and the sounding lead was let down to the depth of five fathoms, without striking the bottom. This river is, perhaps, a branch of the Mackenzie, and falls into a bay, on which I have bestowed the name of my esteemed friend Copland Hutchinson, Esq. Surgeon Extraordinary to His Royal Highness the Duke of Clarence. On its east side there is an island, which was named after Captain Charles Phillips, R.N. to whom the nautical world is indebted for the double-capstan, and many other important inventions.

At five o'clock in the afternoon, rainy weather setting in, we made for a small island, and mooring the boats as near the beach as we could, covered them up, and landed to prepare supper. The length of the day's voyage was twenty-eight miles and a half. Mr. Kendall named the island in honour of Mr. Atkinson, of Berry-House: it is situated in latitude 69° 55′ N., longitude 130° 43′ W., and is separated from a flat, and occasionally inundated shore, by a narrow creek. It is bounded towards the sea by a bulwark of sand-hills, drifted by the wind to the height of 30 feet. Under their shelter 17 winter-houses have been erected by the natives, besides a large building which from its structure, seemed to be intended for a place of assembly for the tribe. Ooligbuck thought it was a general eating-room, but he was not certain, as his tribe erect no such buildings.

I annex a section and ground plan of one of the largest of the dwelling-houses. The centre (A) is a square of ten feet, having a level flooring, with a post at each corner (D,D) to support the ridge-poles,* on which the roof rests. The recesses (B) are intended for sleeping-places. Their floors have a gentle inclination inwards, and are raised a foot above the central flooring. Their back walls are a foot high, and incline outwards like the back of a chair. The ridge-poles are six feet above the floor, the roof being flat in the centre, and sloping over the recesses. The inside of the building is lined with split-wood, and the outside is strongly but roughly built of logs, the whole being covered with earth. An inclined platform (C) forms the ascent to the door, which is in the middle of one of the recesses, and is four feet high; and the threshold, being on a level with the central flooring, is raised three feet above the surrounding ground, to guard against inundations. There is a square hole in the roof, near the door, intended for ventilation, or for an occasional entrance. As we observed no fire-places in these dwellings, it is probable that they are heated, and the cookery performed in the winter, with lamps. Some of the houses were built front to front, with a very narrow passage between them leading to the doors, which were opposite to each other. This passage must form a snug porch in the winter when it is covered with slabs of frozen snow, and one end stopped up. Some of the larger houses which stood single, had log-porches to shelter their doors; and near each house there was a square or oblong pit, four feet

* The ridge-poles were omitted in the section by mistake.

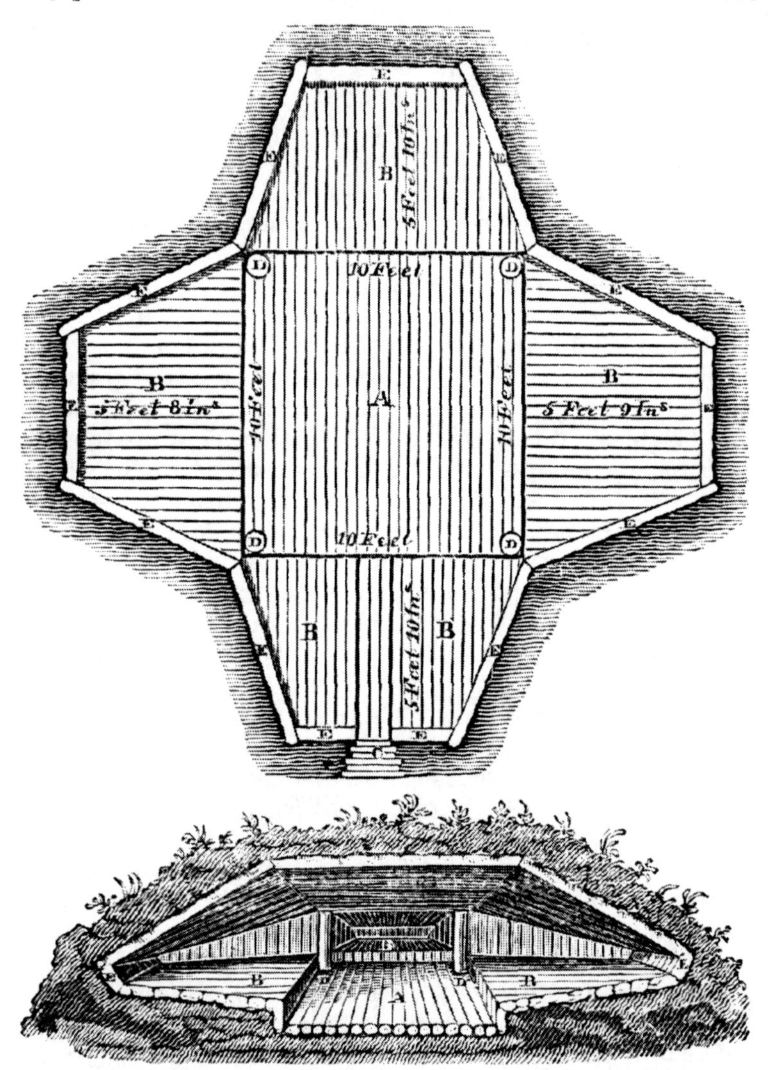

beneath the surface of the ground, lined and covered with drift timber, which was evidently intended for a store-house.

The large building for an assembly-room was, in the interior, a square of 27 feet, having the log-roof supported on two strong ridge poles, two feet apart, and resting on four upright posts. The floor in the centre, formed of split logs, dressed and laid with great care, was surrounded by a raised border about three feet wide, which was, no doubt, meant for seats. The walls, three feet high, were inclined outwards, for the convenience of leaning the back against them, and the ascent to the door, which was on the south side was formed of logs. The outside, covered with earth, had nearly a hemispherical form, and round its base there were ranged the skulls of 21 whales. There was a square hole in the roof, and the central log of the floor had a basin-shaped cavity, one foot in diameter, which was, perhaps, intended for a lamp. The general attention to comfort in the construction of the village, and the erection of a building of such magnitude, requiring a union of purpose in a considerable number of people, are evidences of no small progress towards civilization. Whale skulls were confined to the large building, and to one of the dwelling-houses, which had 3 or 4 placed round it. Many wooden trays, and hand-barrows for carrying whale blubber, were lying on the ground, most of them in a state of decay.

Myriads of musquitoes, which reposed among the grass, rose in clouds when disturbed, and gave us much annoyance. Many snow birds were hatching on the Point, and we saw swans, Canada geese, eider, king, arctic, and surf ducks; several glaucous, silvery, black-headed, and ivory gulls, together with terns and northern divers. Some laughing geese passed to the northward in the evening, which may be considered as a sure indication of land in that direction. The sea-water at Atkinson Island being quite salt, and the ponds on the shore brackish, we had recourse to the ice that lay aground for a supply of fresh water. Strong gales of wind, with heavy rain, continued all night.

Friday, 14th. The rain ceasing at four o'clock in the morning of the 14th, we embarked, and pulled along a sandy bar which projected five or six miles from Atkinson Island, and was covered by masses of ice. We had not left the beach above an hour, when a thick fog hid the land from our view, and a noise of breakers being at the same time heard, we deemed it prudent to moor the boats to a piece of grounded-ice, and wait for clear weather. After a time, the fog dispersing partially, we made sail before a fresh breeze towards the most easterly point of land in sight, but we had not advanced above five or six miles before the looming of the shore on the larboard bow made it necessary to haul to the wind; and the fog becoming as dense as ever, we ran aground on some flats, where the surf nearly filled the boats. On lowering the sails, deeper water was attained, but the wind began to blow hard directly upon the shore, and we could not discover a landing-place, nor did we even know our distance from the beach. In this dilemma we saw a long line of floating sea-weed, and Ooligbuck suggesting that it came from the mouth of a river, we followed its direction, and, with the aid of the sounding lead, groped our way betwixt two shoals into a well sheltered inlet. Here there was a good landing-place, and we deemed ourselves peculiarly fortunate in reaching so snug a harbour, for the fog continued all day, and the wind increased to a heavy gale.

The inlet was named Browell Cove, in honour of the Lieutenant-Governor of the Royal Hospital at Greenwich, and the bay to the westward of it, M'Kinley Bay, out of respect to Captain George K'Kinley, of the Naval Asylum. The latitude of the mouth of Browell Cove is 70° N., and the longitude 130° 19′ W. We did not ascertain its extent, but as its water is brackish, it probably communicates with Esquimaux Lake, which, according to Indian report, lies behind the islands that form this part of the coast. Several large basins of salt water communicate with the cove. Some herds of deer were seen, but too many hunters going in pursuit of them they were frightened away. The temperature throughout the day was 42°.

I observed forty species of plants in flower here, of which nearly one-third were grasses and carices. The Thrift common on the sandy parts of the British coast is a frequent ornament of Browell Cove; and seven or eight of the other plants seen there, are natives of the Scottish hills. Two dwarf species of willows were the only shrubs.

Saturday, 15th.

The fog clearing away, and the wind moderating, we embarked about three in the morning of the 15th; and steering along the coast, came to a group of low sandy islands, that were separated by wide but very shallow channels, and skirted, to the distance of five or six miles, by sand-banks, which were nearly dry at low water. In rounding these banks our soundings varied from two feet to two fathoms, and we were occasionally led almost out of sight of the land. During the whole day we saw much ice to seaward, and in some places it was so closely packed as to render it doubtful whether a ship would have been able to make way through it. The line of deep water was marked by large masses of ice lying aground, and was about ten miles from the shore. As we could not reach the beach, we disembarked upon a piece of ice at noon, and cutting up a spare seat for firewood, proceeded to cook our breakfast, and make observations for latitude and magnetic variation.

After rounding the shoals, we made a traverse of ten miles across an inlet, where the water ran out with a strong current; and, though five fathoms deep, it was nearly fresh. This I supposed to be another communication betwixt Esquimaux Lake and the sea, and named it Russel Inlet, after the distinguished Professor of Clinical Surgery in the University of Edinburgh. The land on its western side was called Cape Brown, out of respect to the eminent botanist, whose scientific researches reflect so much credit on British talent; and that to the eastward of the inlet received the name of Dalhousie, in honour of His Excellency the Governor-in-Chief of the Canadas. Cape Dalhousie consists of a number of high, sandy islands, resembling those seen from Sacred Island, in the mouth of the Mackenzie. We entered some deep inlets amongst them, in search of a landing-place, but the beach was every where too flat. At length, after dragging the boats through the mud for a considerable way, and carrying the cargoes for a quarter of a mile over a flat sand, we reached the shore, and pitched the tents. The island on which we encamped was similar to the others, being from one hundred to one hundred and fifty feet high above the water, and bounded on all sides by steep, sandy cliffs, which were skirted by flat sands. From the summit of the island we had the unpleasant view of a sea covered with floating ice, as far as the eye could reach to the eastward. Temperature during the greater part of the day 55°; at nine P.M. 52°. Wind easterly. The length of this day's voyage was thirty miles and a half; the latitude of the encampment 70° 12′, and longitude 129° 21′ W.

Sunday, 16th. On the 16th the boats were afloat, and loaded by seven in the morning, when we pulled round Cape Dalhousie, and found the land trending as we wished to the south-east. Since reaching the sea, the coast had gradually inclined to the northward, which with the increased quantity of ice seen on the two or three last days, led us to fear that a cape might exist, extending so far to the northward, as to prevent us from reaching the Coppermine River within the period to which our voyage was limited. It was, therefore, with peculiar satisfaction, that, on putting ashore to cook breakfast, we saw distant land to the S.E., apparently of greater height than that which we had recently coasted; and we now flattered ourselves that we were about to leave behind us the low coasts and shoals, which render the boat navigation across the mouths of the Mackenzie and Esquimaux Lake so perplexing and hazardous. Many deer were seen at our breakfasting-place, and the musquitoes annoyed them so much that there would have been no difficulty in approaching them, if we could have spared time to send out the hunters.

Having obtained an observation for latitude, we directed our course to a projecting point across an inlet, with no land visible towards its bottom. The soundings in the middle of the opening exceeded nine fathoms; the water became less salt as we advanced, and at last could only be termed brackish. The point proved to be an Island sixteen miles distant from our breakfasting-place; and as we approached it, we had the mortification to perceive a coast seven or eight miles beyond it, apparently continuous, and trending away to the north-north-west. The island was named Nicholson Island, as a mark of my esteem for William Nicholson, Esq., of Rochester. It is bounded by high cliffs of sand and mud, and rises in the interior to the height of four hundred feet above the sea. The cliffs were thawed to the depth of three feet, but frozen underneath, and the water issuing from the thawing ground caused the mud to boil out and flow down the banks. There were many small lakes on the island, and a tolerably good vegetation. Amongst other plants I gathered here a very beautiful American cowslip, (*dodecatheon,*) which grew in the moist valleys. From the summit of the island a piece of water, resembling a large river, and bearing south, was seen winding through a country pleasantly varied by gently swelling hills and dales, and differing so much in character from the alluvial islands we had just left, that I thought myself justified in considering it to be part of the main land. From S.W. to W.N.W. open water was seen, broken only by a few islands, that were named after Major-General Campbell, of the Royal Marines. This large sheet of water is undoubtedly the Esquimaux Lake, which, according to the natives, not only communicates with the eastern branch of the Mackenzie, but receives, besides, two large rivers; and, consequently, the whole of the land which we coasted from Point Encounter, is a collection of islands. The temperature varied this day from 38° to 55°. The length of the day's voyage was thirty-three miles, the latitude of our encampment 69° 57′, and longitude 128° 18′ W.

Monday, 17th.

On the 17th a thick fog detained us until nine o'clock in the morning, when it dispersed, and we left our encampment. About two miles from Nicholson's Island the water was nine fathoms deep, and had a brackish taste; but as we continued our course to the northward, it became shoaler and salter. This added to the probability of the winding channel, which bore south, being a large river; and that opinion was further strengthened by our observing, when we landed to breakfast, the shore to be strewed with tide-wrack, resembling that which is generally found on the banks of rivers in this country, such as pieces of willows, fragments of fresh-water plants, and lumps of peat earth. We were delighted to find here a beach of sand and fine gravel, bold enough to admit of our running the boats upon it. The fresh footsteps of a party of Esquimaux were seen on the sand.

After obtaining an observation for latitude, we embarked, and continued our course along the coast until we came to the extremity of a cape, which was formed by an island separated from the main by a shallow channel. The cliffs of this island were about forty feet high, and the snow which had accumulated under them in the winter, was not yet dissolved, but, owing to the infiltration and freezing of water, now formed an inclined bank of ice, nearly two-thirds of the height of the cliff. This bank, or iceberg, being undermined by the action of the waves, maintained its position only by its adhesion to the frozen cliffs behind it. In some places large masses had broken off and floated away, whilst in others the currents of melting snow floating from the flat land above, had covered the ice with a thick coating of earth; so that at first sight it appeared as if the bank had broken down; the real structure of the iceberg being perceptible only where rents existed. In a similar manner the frozen banks, or icebergs, covered with earth, mentioned by Lieutenant Kotzebue, in his voyage to Behring Straits, might have been formed. Had the whole mass of frozen snow broken off from this bank, an icebeeg would have been produced thirty feet wide at its base, and covered on one side to the depth of a foot, or more, with black earth. The island was composed of sand and slaty clay, into which the thaw had not penetrated above a foot. The ravines were lined with fragments of compact white limestone, and a few dwarf-birches and willows grew on their sides. The sun's rays were very powerful this day, and the heat was oppressive, even while sitting at rest in the boat; the temperature of the air at noon being, in the shade, 62°, and that of the surface water, where the soundings were three fathoms, 55°.

Immediately after rounding the cape, which was named after His Excellency Sir Peregrine Maitland, Lieutenant-Governor of Upper Canada, we entered a channel ten miles wide, running to the eastward, with an open horizon in that direction; and a doubt arose as to whether it was a strait, or merely a bay. Many large masses of ice were floating in it, which proved to us that it had considerable depth; but the water being only brackish, excited a suspicion that there was no passage through it. While we were hesitating whether to hazard a loss of time by exploring the opening, or to cross over at once to the northern land, several deer were seen, and the hope of procuring a supply of fresh meat, induced us to put ashore and encamp for the night, that the hunters might go in chase. The beach here was strewed with fragments of dark-red sandstone, white sandstone, white compact limestone, and a few pieces of syenite. There were many large trunks of spruce-firs lying on the sand, completely denuded of their bark and branches; and numerous exuviæ of a marine crustaceous animal (*gammarus borealis*) lay at high water mark. Our hunters were successful, Ooligbuck and M'Leay each killing a deer. Many of these animals had fled to the cool moist sands on the coast, but even there the musquitoes tormented them so much as to render them regardless of the approach of the hunters. The latitude of our encampment was 70° 7', longitude 127° 45'; and the length of the day's voyage twenty-three miles. The temperature varied from 52° to 63°. By watching the motion of the tide for the greater part of the night, I fully satisfied myself that the ebb set out of the opening, and that the flood came round the land on the north sine; hence I concluded that there could be no passage to the eastward in this direction, and that the opening led into a bay, to which the name of Harrowby was given, in honour of the Right Honourable the Earl of Harrowby.

Tuesday, 18th.

Embarking on the 18th at three in the morning, we set the sails to a favourable though light breeze, and using the oars at the same time, crossed Harrowby Bay, at its mouth. During the traverse, land was seen round the bottom of the bay. On nearing the shore we distinguished twelve Esquimaux tents on an eminence; and a woman who was walking on the beach gave the alarm, but not until we were near enough to speak to her, her surprise having fixed her to the spot for a time. The men then rushed out, brandishing their knives, and, using the most threatening expressions, forbade us to land, and desired us to return by the way we came. Ooligbuck endeavoured to calm their fears, by telling them that we were friends, but they replied only by repeating their threats, and by hideous grimaces and gestures, which displayed great agility; frequently standing on one foot and throwing the other nearly as high as the head. At length on my bawling "*noowœrlawgo*," (I wish to barter,) they became quiet at once, and one of them running to his kaiyack, and paddling off to us, was followed by many of the others, even before they could witness the reception we gave him. They came boldly alongside, and exchanged their spears, arrows, bows, and some pieces of well-dressed seal-skin, for bits of old iron-hoop, files and beads. They were not so well furnished with iron-work as the Esquimaux we had seen further to the westward, and very eagerly received a supply from us. In our intercourse with them we experienced much advantage from a simple contrivance suggested by Mr. Kendall, and constructed during our halt in Refuge Cove: it was a barricade formed by raising the masts and spare oars eighteen inches above the gunwale on two crutches or davits, which not only prevented our Esquimaux visitors from stealing out of the boats, but, in the event of a quarrel, could have been rendered arrow proof by throwing the blankets or sails over it. On a light breeze springing up we set the sails, and continuing to ply the oars, advanced at the rate of four miles an hour, attended by eleven kaiyacks. Three oomiacks with the women followed us, and we found that, when rowed by two women, and steered by a third, they surpassed our boats in speed.

The females, unlike those of the Indian tribes, had much handsomer features than the men; and one young woman of the party would have been deemed pretty even in Europe. Our presents seemed to render them perfectly happy, and they danced with such ecstacy in their slender boats as to incur, more than once, great hazard of being overset. A bundle of strings of beads being thrown into an oomiack, it was caught by an old woman, who hugged the treasure to her breast with the strongest expression of rapture, while another elderly dame, who had stretched out her arms in vain, became the very picture of despair. On my explaining, however that the present was for the whole, an amicable division instantly took place; and to show their gratitude, they sang a song to a pleasing air, keeping time with their oars. They gave us many pressing invitations to pass the night at their tents, in which they were joined by the men; and to excite our liberality the mothers drew their children out of their wide boots, where they are accustomed to carry them naked, and holding them up begged beads for them. Their entreaties were, for a time, successful; but being desirous of getting clear of our visiters before breakfast-time, we at length told them that our stock was exhausted, and they took leave.

These Esquimaux were as inquisitive as the others we had seen respecting our names, and were very desirous of teaching us the true pronunciation of theirs. They informed us that they had seen Indians, and had heard of white people, but had never seen any before. My giving a little deer's meat to one of them in exchange for fish, led to an inquiry as to how we killed the animal. On which Ooligbuck showed them his gun, and obtaining permission, fired it off after cautioning them not to be alarmed. The report astonished them much, and an echo from some neighbouring pieces of ice made them think that the ball had struck the shore, then upwards of a mile distant. The women had left us previously; several of the men departed the instant they heard the report; and the rest, in a short time, followed their example. They applied to the gun the same name they give to their harpoons for killing whales.

We learned from these people that the shore we were now coasting was part of the main land, and that some land to the northward, which appeared soon after we had passed their tents, consisted of two islands; between which and the main shore, there was a passage leading to the open sea. On landing to cook breakfast and obtain a meridian observation for latitude, we observed the interior of the country to be similar to that seen from Nicholson's Island. The soil was in some spots sandy, but, generally, it consisted of a tenacious clay which cracks in the sun. The air was perfumed by numerous tufts of a beautiful phlox, and of a still handsomer and very fragrant cruciform flower, of a genus hitherto undescribed.

On re-embarking we pulled about eight miles farther betwixt the islands and the main, and found a narrow opening to the sea nearly barred up. The bottom was so soft and muddy that the poles sunk deep into it, and we could not carry the cargo ashore to lighten the boats. We succeeded, however, in getting through, after much labour, and the moment we crossed the bar, the water was greenish, and perfectly salt. The cape forming the eastern point of this entrance lies in latitude 70° 36′ N., longitude 127° 35′ W. and proved to be the most northerly part of the main shore which we saw during the voyage. It is a few miles farther north than Return Reef of Captain Franklin, and is most probably, with the exception of the land near Icy Cape, since discovered by Captain Beechey, in the Blossom, the most northern point of the American Continent. It was called Cape Bathurst, in honour of the Right Honourable the Earl of Bathurst, and the islands lying off it were named after George Baillie, Esq., of the Colonial Office. I could not account in any other way for the comparative freshness of the sheet of water we had left, than by supposing that a sand-bank extended from Cape Dalhousie to Baillie's Islands, impeding the communication with the sea, and this notion was supported by a line of heavy ice which was seen both from Cape Bathurst and Cape Dalhousie, in the direction of the supposed bar, and apparently aground.

Taking for granted that the accounts we received from the natives were (as our own observations led us to believe) correct, Esquimaux Lake is a very extensive and curious piece of water. The Indians say that it reaches to within four days' march of Fort Good Hope; and the Esquimaux informed us that it extends from Point Encounter to Cape Bathurst, thus ascribing to it an extent from north to south of more than one hundred and forty miles, and from east to west of one hundred and fifty. It is reported to be full of islands, to be every where brackish; and, besides its communication with the eastern branch of the Mackenzie, to receive two other large rivers. If a conjecture may be hazarded about the original formation of a lake which we had so few opportunities of examining, it seems probable that the alluvial matters brought down by the Mackenzie, and other rivers, have gradually formed a barrier of islands and shoals, which, by preventing the free access of the tide, enables the fresh water to maintain the predominance behind it. The action of the waves of the sea has a tendency to increase the height of the barrier, while the currents of the rivers and ebb-tide preserve the depth of the lake. A great formation of wood-coal will, I doubt not, be ultimately formed by the immense quantities of drift-timber annually deposited on the borders of Esquimaux Lake.

CONTINUATION OF THE PROCEEDINGS OF THE EASTERN DETACHMENT.

CHAPTER III.

Double Cape Bathurst—Whales—Bituminous-shale Cliffs on Fire—Enter Franklin Bay—Heavy Gale—Peninsula of Cape Parry—Perforated Rock—Detention at Cape Lyon by Wind—Force of an Esquimaux Arrow—Meet with heavy Ice—Pass Union and Dolphin Straits—Double Cape Krusenstern, and enter George the Fourth's Coronation Gulph—Reach the Coppermine River—Remarks—Meteorological Table.

As soon as we entered the clear green water off Cape Bathurst, we perceived a strong flood tide setting against us, and saw several white whales, and some black ones of a large size, but of a species unknown to Ooligbuck.* The natives term them *aggeewœrk*, which is the name given, by the Esquimaux of Hudson's Bay, to the black whales that frequent the Welcome. Many large masses of ice were floating about, but they were no impediment to the boats. The beach, from the time we left Esquimaux Lake, was bold, there being two or three fathoms water close to the shore. We hailed this change of circumstances with pleasure, for the shoals and islands skirting Esquimaux Lake had embarrassed us much, and the brackishness of the water, combined with the trending of the coast

* The appearance of whales on the north coast, nearly midway between the nearest passages into Behring's and Barrow's Straits, and upwards of a thousand miles distant from either, affords subject for interesting speculation. It is known that they must come frequently to the surface to breathe, and the following questions naturally arise :—Are there at all seasons large spaces of open water in the Arctic Seas? or do these animals travel from the Atlantic or Pacific Oceans immediately on the breaking up of the ice off Cape Bathurst, and so early in the season as the middle of July; while the sea, to the eastward and westward, is still covered with ice? If the latter is the fact, it is a very curious part of the natural history of these animals. The Esquimaux informed us, that they are rarely seen when the ice lies close, and in accordance with this remark Captain Franklin saw few to the westward, and we also lost them as we approached the Coppermine River, and met with more ice.

to the northward, and even westward, had excited in our minds an apprehension, that we might possibly be obliged to make a great circuit in search of a passage, out of that extraordinary piece of water, and that the opening, when found, might lie so far to the northward as to be obstructed by an icy sea. Fortunately our fears were groundless; and, to increase our joy, the coast-line from Cape Bathurst appeared to run in a straight direction for Coppermine River. There were many winter-houses built by the Esquimaux on Cape Bathurst. The cliffs facing the sea were still frozen, but the water trickling down their sides showed that they were thawing rapidly. We encamped on the beach in latitude 70° 32½′ N., longitude 127° 21′ W., having sailed that day thirty-seven miles. A plentiful supply of very fine sorrel (*oxyria reniformis*) being obtained from the banks, proved an agreeable addition to our supper.

Wednesday, 19th. Embarking at four o'clock in the morning of the 19th, we rowed along the coast close to the beach, in from two to three fathoms water. We landed at noon to observe the latitude; and at four P.M. a thunder-storm coming on, induced us to encamp for the night. The day's voyage was thirty-two miles, and our encampment was situated in latitude 70° 11′ N., longitude 126° 15′ W., on a point which was named after Dr. Fitton, the distinguished President of the Geological Society. No land was visible to seaward, nor were any fields of ice or large floes seen, but we passed many smaller pieces and some masses, that, having stranded on the beach, were dissolving with great rapidity. A regular tide of six hours affecting the rate of our progress, an allowance was made for it in the reckoning.

The coast consists of precipitous banks, similar in structure to the bituminous-shale cliffs at Whitby, in Yorkshire. They gradually increase in altitude from Cape Bathurst, and near our encampment their height exceeded two hundred and fifty feet. The shale was in a state of ignition in many places, and the hot sulphureous airs from the land were strongly contrasted with cold sea-breezes with which, in the morning, they alternated. The combustion had proceeded to a considerable extent on the point where we landed at noon. Much alum had formed, and the baked clays of yellow, brown, white, and red colours, caused the place to resemble a brick-field or a pottery. This point, which was named after Dr. Traill, of Liverpool, lies in latitude 70° 19′ N. The interior of the country, as seen from the top of the cliffs, appeared to be nearly level, and to abound in small lakes. The soil was clayey, and from the recent thaw wet and soft. Tufts of the beautiful phlox, before mentioned, were scattered over these, otherwise unsightly wastes; and, notwithstanding the scanty vegetation, rein-deer were numerous. Some of the young ones, to whom man was doubtless a novel object, came trotting up to gratify their curiosity, and were suffered to depart unmolested. The sea here abounds in molluscæ, and many black whales were seen; also king-ducks, eiders, snow-birds, hawks, and a large moth.

Thursday, 20th.

We embarked at half-past two on the morning of the 20th, and ran alongshore for two hours with a strong and favourable breeze, when shoals lying off the mouth of a pretty large river, led us six or seven miles from the coast. The breeze, which was off the land, freshened considerably, and raised a short breaking sea, through which we attempted to pull towards the shore, but the boats shipped much water, and made little head-way. We, therefore, set the sails again, and, fortunately fetched under a head-land, and effected a landing. The whole of the pemmican in the Union, and some of that in the Dolphin, was wet on this occasion. In the morning we had passed two Esquimaux tents, pitched on the beach, but the inmates seeming to be asleep, we did not disturb them, being unwilling to lose the fair wind by any delay.

Soon after landing the weather became very foggy, and the wind increased to a heavy gale. The cliffs at our encampment consisted of slate-clay, and bituminous alum-slate, and were six hundred feet high. The river, whose mouth we passed, ran close behind them, having a course parallel to the coast for some miles before it makes its way to the sea. It was named Wilmot Horton River, in honour of the Under Secretary of State for the Colonial Department. Its breadth is about three hundred yards, and it seems, from the quantity of drift-timber that was piled on the shoals at its mouth, to flow through a wooded country. The length of this day's voyage was twenty-four miles, and the position of our encampment was in latitude 69° 50′ N., longitude 125° 55′ W. At high-water, which took place at a quarter past four in the afternoon, the small slip of beach on which we had encamped was almost covered, and we had to pile the baggage on the shelving cliff. A very showy species of gromwell grew near our encampment, in company with the common sea-gromwell, (*lithospermum maritimum.*)

Friday, 21st.

On the 21st strong winds and foggy weather, with a considerable surf on the beach, detained us until after eight o'clock in the morning, when many large masses of ice coming in, took the ground near the shore, and smoothed the water sufficiently to enable us to embark. The fog was dense to seaward and over the land, but the height of the cliffs left a space of about a mile from the beach, over which it was carried by the violence of the wind.

About two miles from our late encampment, the bituminous shale was again noticed to be on fire, giving out much smoke; and as we advanced, the cliffs became less precipitous, appearing as if they had fallen down from the consumption of the combustible strata. They gradually terminated in a green and sloping bank, whose summit, about two miles from the sea, rose to the height of about six hundred feet. For the information of the general reader, I may mention that the shale takes fire in consequence of its containing a considerable quantity of sulphur in a state of such minute division, that it very readily attracts oxygen from the atmosphere, and inflames. The combustion is rendered more lively by the presence of bitumen; and the sulphuric acid, which is one of its products, unites with the alumina of the shale to form, with the addition of a small quantity of potass, the triple salt, well known by the name of alum. The moistening of the strata by the sea-spray accelerates the process. In some alum-works, where nature has not been so favourable as in the cliffs of Cape Bathurst, a deficiency of the bituminous matter requisite to keep up the proper intensity of combustion, is supplied by brush-wood, which is strewed in alternate layers with shale that has been previously much divided by long exposure to the weather, and the whole is then moistened with salt-water. A further account of these cliffs is given in page xl. of the Appendix.

In the forenoon we passed the mouths of two small rivers, which were designated after Sir Henry Jardine, Bart., King's Remembrancer in the Court of Exchequer for Scotland; and Dr. Burnett, Commissioner of the Victualling Board. A meridian observation was obtained in latitude 69° 36′ N.

In the afternoon the wind blowing more on the shore, caused a tumbling sea. We sailed amongst much stranded ice, and, following the line of coast, were gradually led into a deep bay, whose east side, having a northerly direction, was formed by low land, and so much broken by numerous and extensive inlets, as to look more like a collection of islands than a part of the main land. We were now, reckoning by degrees of longitude, fully half way from Point Separation to the Coppermine River, and the coast from Cape Bathurst had been so exactly in the proper direction, as to excite high hopes of a short and prosperous voyage: it was, therefore, no pleasant sight to us to behold land running out at right angles to our course, and we were willing to believe that a passage existed betwixt it and the main. This opinion was supported by the direction of the high land, which had hitherto skirted the shore, continuing to be south-easterly, until lost to the sight at the distance of fifteen or twenty miles. We, therefore, endeavoured to find a passage, but the first opening that we came to, led into a circular basin of water, apparently land-locked, and about five miles in diameter. We halted at its entrance to cook our supper, and, during our stay, perceiving that the ebb-tide set out of it, we determined on searching for a passage elsewhere. This inlet is six fathoms deep at its entrance, and would prove an excellent harbour for a ship, only for the sand-banks, which skirt this part of the coast, and which render the passage into it too intricate for vessels having a greater draught of water than our boats. It was named Langton Harbour, after the agent for the Hudson's Bay Company at Liverpool.

Leaving this harbour, and steering to the northward, we passed several inlets, into which the flood-tide set with a strong current. We could not see land towards their bottoms, but their mouths were shoal, and we felt convinced that there was no passage through them, because the flood-tide entered them from the westward. We, therefore, proceeded on our voyage without wasting time in examining them; and at two o'clock, on the morning of the 22nd, having come fifty-four miles, we encamped on a beach composed of small fragments of limestone, and strewed with sea-weed. This beach, which received the name of Point Stivens, separates an extensive sheet of salt-water from the sea, and is similar in character to the Chesil Beach, that connects the Isle of Portland to the shore. It varies in breadth from one hundred yards to a quarter of a mile, is several miles long, has a northern direction, and seems to have been formed by the sweep of the tide round the bay, meeting the ebb from the basins that intersects the peninsular promontory with which it is connected. There are several narrow breaches in it through which the tide flows. Anxious to discover the termination of this promontory which was leading us so much out of the direct course to the Coppermine, I went to the summit of a rising ground, about five miles distant, but the view was closed by some small hills, two or three miles off. The soil was clayey, and vegetation scanty.

Saturday, 22nd.

In taking wood to make a fire, from a large pile of drift-timber which had been collected by the Esquimaux, the nest of a snow-bird, containing four young, was discovered. The parent bird was at first scared away, but affection for its offspring at length gave it courage to approach them with food; and as it was not molested it soon became quite fearless, and fed them with the larvæ of insects, whilst the party were seated at breakfast close by the nest.

At nine o'clock, A.M., we embarked again, and running before a favourable breeze, came to a point consisting of cliffs of limestone, twenty feet high, with a small island of the same kind of rock at its extremity. Many large boulders of greenstone were seen here. After ascertaining the latitude by meridian observation to be 69° 42′ N., we continued our voyage along a bold shore, consisting of precipices of limestone, forty or fifty feet high, with three or four fathoms of water at the base. In the evening, having reached a projection which appeared to be the western pitch of the cape, we encamped in a bay near a remarkable perforated rock, having come twenty-six miles since leaving Point Stivens. In the course of the day's voyage we had to make our way through some pretty extensive streams of ice, composed of pieces which rose eight or ten feet above the water; and we saw a considerable quantity of what is termed sailing ice to seaward, being such as a ship could make her way through. I had now the gratification of naming the extensive bay we had been coasting for three days, after my friend and commanding officer; and to the several inlets on its eastern side I assigned the names of Wright, Cracroft, and Sellwood, in honour of his near relatives. A group of islands to the northward was named Booth Islands on the same account.

In bestowing the name of Franklin on this remarkable bay, I paid an appropriate compliment to the officer, under whose orders and by whose arrangements the delineation of all that is known of the northern coast of the American Continent has been effected; with the exception of the parts in the vicinity of Icy Cape discovered by Captain Beechey. It would not be proper, nor is it my intention, to descant on the professional merits of my superior officer; but after having served under Captain Franklin for nearly seven years, in two successive voyages of discovery, I trust I may be allowed to say, that however high his brother officers may rate his courage and talents, either in the ordinary line of his professional duty, or in the field of discovery, the hold he acquires upon the affections of those under his command, by a continued series of the most conciliating attentions to their feelings, and an uniform and unremitting regard to their best interests, is not less conspicuous. I feel that the sentiments of my friends and companions,

Captain Back and Lieutenant Kendall, are in unison with my own, when I affirm, that gratitude and attachment to our late commanding officer will animate our breasts to the latest period of our lives. After this feeble but sincere tribute of respect and regard, in which I hope I have not overstepped the proper bounds of a narrative, I hasten to resume the details of the voyage.

The country in the neighbourhood of the encampment consisted entirely of limestone, mostly of the variety named dolomite, and, as is usual where that stone prevails, it was extremely barren. The cliffs and points of land present many caverns and perforated rocks, which have very strong resemblances to the windows and crypts of Gothic buildings. The common kittiwake breeds in great numbers on the rocky ledges in this quarter, and their young were already fledged. The temperature during the day was nearly stationary at 46°, the wind south. The evening being very fine, the pemmican was taken out of the bags, which were scraped and dried; and our loss of provision, by the wetting it sustained in the gale of the 20th, proved to be less than we had expected.

Sunday, 23rd.

Embarking at four o'clock, A.M. of the 23d, we sailed with a favourable breeze for nine miles, betwixt Booth Islands and a shore presenting alternately projecting rocky shoals and narrow inlets. We then landed, and ascended a hill, about seven hundred feet high, to ascertain the direction of the coast, and had the satisfaction of finding that we had now reached the northern extremity of this remarkable promontory. It was named Cape Parry after the distinguished navigator whose skill and preservance have created an era in the progress of northern discovery, and a letter addressed to him, containing information of our proceedings and of Captain Franklin's as far as was known to us, was deposited under a pile of stones which we erected on the summit of the hill. From this elevated situation, land was faintly seen bearing S.E. by S., about forty miles distant; and from thence round to Booth Islands there appeared an open sea, merely studded with a few streams of sailing ice, but no islands were seen in that direction. There are many well sheltered coves in the vicinity of Cape Parry and amongst Booth Islands, but the bottom is rocky, and numerous reefs render the navigation unsafe for a ship. The eastern side of Cape Parry exhibits a succession of limestone cliffs, similar to those which from its western shores; and as we continued our voyage, we passed many excavations ornamented by graceful slender pillars, and exhibiting so perfect a similarity to the pure Gothic arch, that had Nature made many such displays in the Old world, there would be but one opinion as to the origin of that style of architecture. A small island, on which we landed to cook breakfast, was named after the late Daniel Moore, Esq., of Lincoln's Inn. It was composed of a cellular limestone, containing many crystals of quartz. The whole party went in pursuit of a polar hare which was seen here, but, although it had no other shelter than the rocks, it contrived to escape from us all.

In the evening we encamped on an island, which was named by Mr. Kendall after the Reverend Dr. Burrow of Epping. It is situated in latitude 69° 49′ N., longitude 123° 33′ W. The length of the day's voyage was thirty-one miles. Fine weather, and a temperature of 52°, entailed upon us a visit from the musquitoes. The sea water here is of a light blue colour and clear, the bottom being distinctly visible in five fathoms. Pieces of ice still adhered to the cliffs.

Monday, 24th. We were detained in the morning of the 24th by a thick fog, which cleared up about eight o'clock; but the moon being then in distance, we remained until noon, that Mr. Kendall might take observations for lunars and latitude. These necessary operations being completed, a short voyage of nine miles brought us to an island on which we encamped, and which obtained from us the name of Clapperton, in honour of the undaunted explorer of central Africa. In our way we passed through several streams of ice, composed of pieces of considerable size, but all evidently in a state of rapid dissolution, under a bright sun; the water flowing from their surfaces in rivulets. Many black whales, and various kinds of seals, were seen this day. We saw no black whales farther to the eastward.

From Clapperton Island we had a view of a ridge of hills, which, from their direction, appeared to be a continuation of those on the west side on Franklin Bay. The island itself, like the neighbouring coast, is composed of limestone, and many detached rocks skirt it, rising from water that is beautifully clear. When we landed there was a strong current setting to the eastward, round the end of the island, but it ceased at four P.M., the time of low water, and was probably produced by the ebb setting out of some of the inlets of Cape Parry. In the evening the ice made a noise so like the regular firing of half-minute guns, as to excite, at first, an idea that we heard the guns of a ship. The temperature at six o'clock in the evening was as high as 74° in the shade.

Clapperton Island lies in latitude 69° 41½′ N., and nearly in the longitude of Fort Franklin, from which it is distant three hundred and thirteen miles in a straight line; but the distance between the nearer part of the Great Bear Lake and the Arctic Sea here, does not much exceed one hundred and ninety miles.

Taking advantage of a light breeze and very fine weather, we embarked at midnight, and crossed over to the east side of the bay, passing through some heavy streams of ice by rather intricate channels. At half past five in the morning of the 25th, we landed on a point of the main shore, and Mr. Kendall took observations for three sets of lunars. On re-embarking we proceeded a few miles further, when a heavy gale of wind suddenly springing up, we ran for shelter into a small creek at the extremity of a cape, which I named after the distinguished traveller Captain G. F. Lyon, R.N. The bay which lies betwixt it and Cape Parry, was called Darnley, in honour of the Earl of Darnley. The distance from Clapperton Island to Cape Lyon is fourteen miles.

Tuesday, 25th.

The country in the neighbourhood of Cape Lyon presents a surface varied by gently swelling eminences, covered with a grassy sward, and intersected by several narrow ridges of naked trap rocks, rising about one hundred and fifty feet above the general level. The trap ridges, when they reach the coast, form high cliffs, and the clay-slate and limestone lie in nearly horizontal strata beneath them. The view inland was terminated by the range of hills which we had seen at the bottom of Darnley Bay, to which the name of Melville Range was now given, in honour of the Right Honourable the Lord Viscount Melville.

From the top of the highest trap-hill, near the extremity of the cape, we saw some heavy ice to seaward, but with enough of open water for the passage of a ship; and, occasionally, during our stay, there was an appearance of land to the north-westward, occupying two points of the compass; but we were uncertain whether it might not be a fog-bank hanging over a field of ice. If it was land, it could not be less than twenty-five or thirty miles distant, and must, from the portion of the horizon it occupied, be a large island. Upon the summit of the hill we erected a pile of stones, and deposited another letter for Captain Parry, containing a short account of our proceedings.

A gale of wind detained us two days at Cape Lyon, during which Ooligbuck supplied us with rein-deer meat, and Mr. Kendall obtained several sets of lunars. The latitude of our encampment, by the mean of three meridian observations, was 69° 46½′ N.; and the longitude, by lunar distances, 122° 51′ W.

The temperature of the air, during the gale, was about 45°, that of the water 35°. During our stay at Cape Lyon the tides were regular, but the rise and fall were short of twenty inches. At midnight on the 26th of July, the sun's lower limb was observed to touch the horizon for the first time since our arrival on the coast. Some old winter houses were seen in our walks, but we perceived no indications of the Esquimaux having recently visited this quarter.

Thursday, 27th. The gale moderated on the 27th, and at eight in the evening it was sufficiently abated to permit us to proceed on our voyage. After rowing about two miles, the horns of a deer were seen over a rock at the summit of a cliff, on which M'Leay, the coxswain of the Union, landed and killed it. This poor animal had been previously wounded by an Esquimaux arrow, which had broken its shoulder bone. The jagged bone-head of the arrow was buried in the flesh, and its copper point bent up where it had struck the bone. The wound was open, and seemed to have been inflicted at least a fortnight before, but the animal was still fat. The extremity of Cape Lyon lies about three miles northeast of the encampment we had left, and in its neighbourhood the cliffs form bold head-lands and several small rocky islands. Soon after rounding it we came to a projecting point, consisting of cliffs of limestone, in which there was a remarkable cave, opening to the sea by an archway, fifty feet high and twenty wide. The walls of the cavern were two hundred feet high, and a large circular aperture in the roof gave free admission to the day-light. Mr. Kendall named this point after Mr. Pearce, a particular friend of his.

The night was fine but cold, the temperature having fallen to 35° soon after we started, and at midnight the sun sunk for nearly half an hour beneath the horizon. We passed much heavy stream-ice, and towards the morning a quantity of new, or, as the seamen term it, "bay ice," having formed on the surface on the sea, the boats were so much retarded that we put ashore at four o'clock of the 28th, to wait until the increasing heat of the day dissolved it. The point on which we landed was named after Admiral Sir Richard Godwin Keats, G.C.B., Governor of Greenwich Hospital, and lies in latitude 69° 49′ N., and longitude 122° W., being about eighteen miles distant from our encampment on Cape Lyon. The rocks at Point Keats consist of flesh-coloured sandstone. The Melville range of hills approaches there within eight or ten miles of the sea, and the intervening country is traversed by ridges of greenstone. On the coast from Cape

Friday, 28th.

Lyon to Point Keats there is a line of large drift timber, evidently thrown up by the waves, about twelve feet perpendicular height, above the ordinary spring-tides: a sufficient proof of the sea being nearly clear of ice at the time it was thrown up; for the presence of any considerable quantity, even of stream-ice, prevents the waves from rising high. After two hours halt, the bay-ice having dissolved we re-embarked.

From Cape Lyon to Point Keats the coast runs nearly east; after quitting the latter we found it trending a little to the southward, and from a point, which was named in honour of John Deas Thompson, Esq., Commissioner of His Majesty's Navy, it has nearly a south-east direction, We landed a little to the eastward of Point Deas Thompson, to take a meridian observation for latitude, in a small bay, bounded by cliffs of limestone, one hundred and forty feet high, in which the waves had sculptured some beautiful Gothic arches. From the summit of the cliffs, we saw a dark appearance in the eastern horizon, but it was too indistinct to permit us to decide whether it was land or merely a fog-bank. To the eastward of these cliffs the coast decreased in height, and, at the distance of five miles, we passed a small river, which was named after Francis Palgrave, Esq. Near this river, on the summit of a cliff, which was twenty-five feet high, we noticed several large logs of drift timber, with some hummocks of gravel, that appeared to have been thrown up by the waves. A portion of the Melville Range lies within three miles of the shore there; and one of its most remarkable hills was named after my esteemed friend, William Jackson Hooker, LL.D., Regius Professor of Botany in the University of Glasgow; and another after Colonel Colby, of the Royal Engineers, one of the Members of the Board of Longitude. About four o'clock in the afternoon we came to a stream flowing from a lake, and as it was an excellent boat harbour, we entered it and encamped. It was named Roscoe, after the eloquent historian of the Medici; and a conical hill of the Melville Range, visible from its mouth, received the name of the venerable geographer Major Rennel.

We passed this day through heavier and more crowded streams of ice than any we had previously seen on the voyage. The navigation amongst it was tedious and difficult, and just before we put ashore much motion was imparted to it by a fresh south-west wind. The temperature during the day varied from 35° to 50°. The mouth of Roscoe River lies in latitude 69° 41′ N., longitude 121° 2′ W., and is forty-eight miles distant from Cape Lyon.

Saturday, 29th. We embarked on the 29th, with a fair wind; but the ice lay so close, that we could not venture to set more than a reefed foresail, and were ultimately obliged to lower the sail entirely, and to find a passage through ice with oars and poles. The pieces of ice were of sufficient magnitude to deserve the name of floes, and were sometimes several fathoms thick. They were all moving before the breeze, which caused them to arrange themselves in the form of streams parallel to the coast, and, consequently, left lanes of open water in the direction of our course. These lanes, however, were continually changing their form; and, on several occasions, when we had been tempted by the favourable appearance of a piece of open water to venture from the coast, we had great difficulty in extricating ourselves from the ice which closed around us. The thickness of the ice led me to conclude that the sea had not been long open in this quarter; and I observed that the vegetation was later on this part of the coast than on the western side of Cape Parry.

For the first twelve miles after leaving our encampment, the coast was low and sandy; the Melville Range still forming the back-ground, at the distance of four or five miles from the sea. The low beaches were terminated by a rocky headland, which obtained from us the name of De Witt Clinton, as a testimony of our sense of the urbanity and love of science which had prompted his Excellency the Governor of the state of New York* to show so much attention to the members of the Expedition, in their passage through his government. Some miles beyond Point De Witt Clinton we came to a steep cliff, where the ice was so closely packed that we could not force a passage. The cargoes were, therefore, carried along the foot of the cliff, and the boats launched for a few yards over a piece of ice. In this operation, the shelving base of an iceberg, which had formed under the cliff, and still adhered to it, but which was undermined by the waves, gave way whilst several of the men were standing upon it; but, fortunately, it did not overset, and they received no injury, as it was large enough to support them in the water. At nine o'clock, A.M., we were stopped by the closenes of the ice, and put ashore until the tide or wind should produce some change.

The tides, since leaving the Mackenzie, had never been observed to have a greater rise than eighteen inches: but in the neighbourhood of our encampment, the sea-wrack and lines

* Since the above passage was written, the world has had to mourn the loss of this distinguished statesman and philosopher.

of drift timber indicated a washing of the sea to the perpendicular height of twenty feet. The country in this vicinity consists of a bluish limestone, interstratified with slate-clay: and naked and rugged ridges of trap rocks rise in various places above the general level. The soil is composed of clay and limestone gravel. The latitude of our encampment was ascertained, by meridian observation, to be 69° 29′ N.; its longitude was 120° 20′ W.; and its distance from Roscoe River was twenty-five miles.

Sunday, 30th.

A breeze of wind from the land having opened a passage two miles wide, we embarked at two o'clock in the morning of the 30th, and ran seven miles under sail; when, having overtaken the ice which had passed in the night, we found it too closely packed to allow us to proceed. In making for the beach, the Union narrowly escaped being crushed by two large floes of ice, which came together with violence just as she was about to run betwixt them. The Dolphin had sailed through the same passage not two minutes before. From an eminence near our encampment, we had the unpleasant view of a sea covered, as far as the eye could reach, with ice, excepting a few lanes of open water far to seaward. The tide fell here seven inches in the morning, and eleven in the evening, although the north-west wind increased in the afternoon to a pretty strong gale. The greater fall of the water with that wind, showing that it found an exit to the eastward, relieved us from an apprehension, which we had begun to entertain, that we were entering a deep bay, which might be encumbered by the drift-ice for many days. Much ice drove past us in the course of the day, before a west-north-west wind, its progress being only slightly checked for a time by the flood tide. Recent footsteps of a small party of Esquimaux were seen on the beach. Our encampment was situated in latitude 69° 24′ N., and longitude 120° 03′ W.

Monday 31st.

Embarking on the 31st, at two o'clock in the morning, we succeeded in getting about six miles through the ice: when we were again obliged to put ashore at the mouth of a small river, which was named after James Buchanan, Esq., his Majesty's Consul at New York, whose friendly attention to the officers of the Expedition well entitled him to their gratitude. After waiting for awhile the tide loosened the ice a little, and we made some progress by debarking upon the floes, and pushing them apart with poles, until a sufficient opening was made. This operation was tedious, and not devoid of hazard to the boats, arising from the rotatory motion frequently given to the floes, by the pressure of the body of the ice. At noon, an observation for latitude was obtained on a projecting point, which was named after William Tinney, Esq., of Lincoln's Inn. At three in the afternoon, our progress being again arrested by the compactness of the ice, we hauled the boats upon the beach, and M'Leay having killed a fat buck rein-deer, the party had an excellent supper after the fatigues of the day. The length of the day's voyage was twenty-two miles; the latitude of our encampment 69° 17½' N., and its longitude 119° 27' W. The coast line in this quarter is lower, few of the cliffs exceeding forty feet in height, and there is a greater proportion of flat beach than occurs nearer Cape Lyon. The ground is strewed with gravel, apparently arising from a limestone conglomerate which exists there in considerable quantity. The Melville Range is within four or five miles of the shore at this place, and does not rise more than five hundred feet above the sea. Many small rivulets flow from the rising grounds into the sea, through wide gravelly beds, indicating that at times they swell into large torrents.

A light westerly wind having opened a narrow channel between the ice and the shore, we embarked early in the morning of the 1st of August, and, three miles from our encampment, came to a river, which discharged itself by various shallow mouths, separated by sand banks. Its westernmost and easternmost mouths were five miles apart; and the latter, which was the largest, was one hundred and fifty yards wide. Although the outlet of this river is so much barred up, it discharges a considerable volume of water, and probably has its sources in the hills which are visible from the northern shores of Great Bear Lake. It was named after John Wilson Croker, Esq., Secretary to the Admiralty. Further on we had a view of a high island, lying ten or twelve miles from the shore, which received the appellation of Sir George Clerk's Island. M'Leay, who was now acknowledged to be our best hunter, was sent in pursuit of a deer, which we saw from the boats, and being successful, we landed to cook our breakfast, after having rowed twenty miles in the course of the morning. An observation for latitude was obtained a mile further at a point which was named after Waller Clifton, Esq., Secretary to the Victualling Board. The coast here makes a turn to the southward, and about six miles further on, where it resumes its easterly direction, a river about one hundred yards wide flows into the sea, betwixt two sand hills. To this river Mr. Kendall gave the name of Inman, out of respect to the Reverend and Learned Professor of the Royal Naval College

August 1st.

at Portsmouth. A conical hill, about ten miles distant, in a south-west direction, was named after the late President of the Royal Society, the highly distinguished Sir Humphry Davy, Baronet. This was the last part we saw of the Melville Chain. We encamped at half past seven in the evening, under a high cliff of lime-stone, having advanced during the day thirty-seven miles. The point on which we encamped, received from Mr. Kendall, the name of Wise, after Captain M. F. Wise, of the Royal Navy, under whose command he sailed in His Majesty's ship Spartan.. It is situated in latitude 69° 03½′ N., longitude 118° W.

The coast from Cape Clifton to Point Wise consists of lime-stone in horizontal layers, forming cliffs, which are separated from each other by intervening shelving beaches, and it is skirted to the distance of a quarter of a mile by rocky shoals, having sufficient water on them for our boats, but not enough to admit the heavy ice. This was the cause of our making greater progress than we had been led to expect from the appearance of the ice in the morning. The cliffs at Point Wise are two hundred feet high, and from their summits, the ice appeared closely packed, as far as the eye could reach; no lanes of open water being visible. It was, however, composed of pieces, and not a continuous field, for we could distinctly perceive that several of the hummocks it inclosed were in motion. This was the first time during the voyage that we saw ice so closely packed, as to appear impenetrable to a ship when impelled by a good breeze, but it is necessary to state that, even from a considerable height, we could not tell with certainty the state of the ice six miles off; scattered pieces at that distance assuming the appearance of a close pack. The weather this day was fine, the temperature varying from 43° to 50°.

Wednesday, 2d.

Soon after setting out on the 2d, the temperature, which had been about 40° throughout the night, fell to 34°, and a fog came on. The wind also freshening and putting the ice in motion, the boats received some heavy blows; but we continued to advance, though slowly, and with much caution. About ten miles from our encampment, we passed the mouth of a small river, which was named after Captain Hoppner, of the Royal Navy, second in command to Captain Parry, on his third voyage of discovery. Towards noon the fog cleared away, and a meridian observation was obtained in latitude 68° 56′ N. Three miles further on we arrived at the mouth of a wide but shallow river, which flowed over a rocky bottom, betwixt two sand hills, and joined the sea by several mouths, separated by shoals. To this river Mr. Kendall gave the name of his friend, Lieutenant Harding, of the Royal Navy. Five miles beyond this river, on the extremity of a rocky cape, the Esquimaux had constructed several store-houses, of drift timber, which were filled with dried deers-meat and seal-blubber; along with which, cooking kettles, and lamps made of potstone, copper-headed spears, and various other articles, were carefully laid up. The ashes of the recently extinguished fires showed that the natives had quitted this place only a few days, and we felt much pleasure in figuring to ourselves the surprise and joy with which they would behold, on their return, the iron utensils that we deposited in the store-houses for their use. The cape received the name of "Young," after the learned Secretary to the Board of Longitude.

From Cape Young we had a view of the sea thickly covered with ice, of a greater thickness than any we had previously encountered; and we perceived that there was a deeply indented bay lying in our route, and so filled with ice, that our only method of passing it appeared to be by keeping close to the shore, although under the disadvantage of trebling the distance. The coast in this quarter is similar to that which we had passed on the two or three preceding days, and is formed of high limestone cliffs, with intervening shingly beaches; but the country is still more barren, the quantity of limestone debris almost excluding any soil. Flat limestone rocks, having only a few inches of water upon them, skirt the beach, and terminate like a wall in four or five fathoms water. The ice was closely packed against these rocks, and for five miles after passing Cape Young, we made a way for the boats only by the constant use of the hatchet and ice-chisel, and gladly encamped at six o'clock in the evening, after a day's voyage of thirty-one miles. A herd of twenty rein-deer were grazing on the beach, but our hunters were too much fatigued to go in pursuit of them. The encampment was situated in latitude 68° 53′ N., and longitude 116° 50′ W. The temperature varied in the course of the day from 34° to 50°. We observed that the ice continued to dissolve, but not so rapidly as in the month of July, when the sun did not sink below the horizon.

Thursday, 3rd. We resumed our operations on the morning of the 3d at the usual hour, and with great labour made a passage for the boats. At eleven o'clock we landed to refresh ourselves on a projecting point at the western entrance of a deep bay, having previously passed a river which was about one hundred yards wide, but very shallow. After breakfasting, and obtaining a meridian observation in latitude 68° 53′ N., we pushed off again, and for some time made very slow progress. The shores of the bay consisted of beds of limestone, which, shelving into the water, were covered with masses of ice, forced up by the pressure of the pack outside. We were, therefore, compelled to work our way in deeper water, and there the boats, which led by turns, were occasionally exposed to the hazard of being overset by pieces of buoyant ice, which frequently broke off from the bases of the floes. In the language of the whalers, the ice is said to *calf*, when masses are detached in this manner, and they are sometimes of sufficient magnitude in the Greenland seas to endanger large vessels. The Dolphin was, at one time, nearly crushed to pieces by the closing of two floes; but, fortunately, she had reached a small recess, just as they came in contact, and they recoiled sufficiently to leave a passage for her exit, after she had sustained the trifling damage of a few cracks in the upper planks. The rays of the sun, and the waves acting on the surface of the floes, had, by thawing them irregularly, formed lakes of fresh water of some extent upon their surface. When these pieces of water were of sufficient depth, we availed ourselves of them to make some progress in our voyage, and in this way we frequently sailed over a considerable thickness of ice.

At four o'clock P.M. we had advanced five miles, when to our joy we found a lane of open water, which permitted us to cross to the other side of the bay, where we encamped in latitude 68° 51½′ N., and longitude 116° 03′ W., having sailed in the course of the day eighteen miles and a half. The bay was named Stapylton in honour of Major-General the Honourable G. A. C. Stapylton, Chairman of the Victualling Board; and on ascending a rising ground we perceived that it communicates with a long, narrow lake. A few miles from the coast the land rises from three to five hundred feet above the sea, and presents many precipitous limestone cliffs, and chains of small lakes. The country is very barren, the only plant we gathered being the yellow poppy, (*papaver nudicaule.*) By our reckoning we were now nearly in the longitude of the mouth of the Coppermine River, but about seventy miles to the northward of it, we, therefore, entertained an opinion that we were coasting a narrow peninsula, and that we should soon have the pleasure of perceiving the coast take a southerly direction. It was, consequently, with some hopes of beholding the sea on the opposite side of the peninsula that I walked seven or eight miles to the eastward in the night, but

I was disappointed. In my way I had occasion to wade through a small lake, when two birds, about the size of the *northern diver*, and apparently of that genus, swam, with bold and angry gestures, to within a few yards of me, evidently very impatient of any intruder on their domain. Their necks were of a beautiful pale yellow colour, their bodies black with white specks. I considered them to belong to a species not yet described, and regretted that, having left my gun at the tent, it was not in my power to procure one of them for a specimen.

Friday, 4th. Embarking at three A.M. on the 4th, we found little difficulty in reaching the eastern cape of Stapylton Bay, the wind having formed a narrow channel between the ice and the shore in the night. The temperature was low, and in the morning some new ice was formed which we easily broke. We noticed several eider ducks breaking a way through the thin ice for their young ones with their wings, and in this operation they made greater progress than we did in the boats.

On reaching the cape* which was named after Vice-Admiral Sir William Johnstone Hope, G. C. B., we descried another point about four or five leagues distant, bearing east-north-east, the intervening bay being filled with closely packed ice. We were now within twelve miles of Cape Young, after a laborious navigation of four times that distance, and the prospect of another bay, equally unpromising, was very vexatious; but our apprehensions were increased by the view of a continuous line of land, extending from north-north-west until it was hid behind the nearer cape, which bore east-north-east, for we feared that it might prove to be a continuation of the main shore. Our crews, though concerned at the delay that so much ice was likely to occasion, set about overcoming the obstacle with a hearty good will, and after an intricate and troublesome navigation of ten or twelve miles amongst the ice, we found the bottom of the bay more open, and were enabled to cross over to the eastern side where we encamped. This bay received the name of the eminent astronomer James South, Esq.

Mr. Kendall having gone to ascertain from the higher ground the trending of the coast, returned in about two hours with the cheering intelligence that the land to the northward was unconnected with the main shore, and that he had seen the latter inclining to the south-east, with a much more open sea than we had lately been accustomed to. As soon as sup-

* Its latitude was ascertained by meridional observations to be 68° 58′ N.

per was over, I also set out to enjoy the gratifying prospect, and from the extremity of the cape on which we were encamped, and which was named in honour of the Right Honourable Lord Bexley, I beheld the northern land running from north-north-west till it was lost in the horizon on a north 73° east bearing. It seemed to be pretty high but not mountainous; and although broken towards the east, the principal portion of it appeared to be continuous. This island, by far the largest one that was seen, either in the present voyage or on Captain Franklin's former Expedition, was named after that most distinguished philosopher Dr. Hyde Wollaston. The main shore had a direction nearly parallel to Wollaston Land, its most distant point in sight, which I estimated to be fifteen miles off, bearing S. 61° E. On the strait, separating the two shores, I bestowed the names of our excellent little boats, the Dolphin and Union. It varies in width from twelve to twenty miles, and to the eastward semed to contain merely detached streams of ice, not likely to obstruct the progress of a vessel; but to the westward lay the closely packed ice, filling South's Bay, and extending to seaward. The ice did not, however, entirely close the strait, for I could discern lanes of open water towards Wollaston Land. The packed ice which we had seen lining the coast between Point Clifton and Cape Bexley, may be perhaps considered as an illustration of the remark made by Captain Parry, that the western sides of seas and inlets in those latitudes are more encumbered with ice than the opposite sides; and it is very probable that a ship might have found a passage by keeping along Wollaston Land, an opinion which the appearance of the ice as seen from Cape Bexley, tended to confirm. The latitude of our encampment was 68° 58′ N., and its longitude 115° 47′ W.; it was within ten miles of our encampment of the preceding night, although we had travelled twenty-five miles in the course of the day.

Saturday, 5th.

The party embarked on the 5th, at the usual hour in the morning, with their spirits pleasantly excited by the intelligence of the favourable trending of the coast, communicated by Mr. Kendall, and after doubling Cape Bexley, proceeded under sail, before a west-north-west wind, with a rapidity to which they had lately been unaccustomed. The point of land which Cape Bexley terminates, consists entirely of horizontal beds of limestone, and is nowhere more than three hundred feet above the sea. On the west side, the water is two or three fathoms deep, close to the shore, and the land attains its greatest elevation by a steep rise from the beach. On the east side there are some precipitous cliffs, but the coast in general is skirted by shelving rocks. No soil was seen on the Cape, nor any appearance of vegetation, the ground being every where covered, to the depth of a foot, by fragments of limestone, which are detached by the frost from the solid strata lying beneath. We were much puzzled at first with the appearance of several parallel trenches, a foot deep, running for a great distance amongst the fragments, but on examination they were ascertained to originate in fissures of the subjacent strata. Much quartz being intermixed with the limestone of Cape Bexley, the fragments which covered the ground had, by the action of the weather, lost most of the softer calcareous matter, and were converted into a kind of rasp, very annoying to pedestrians, being capable of destroying a pair of stout English shoes in a walk of a few hours.

At eleven o'clock we came to a pack of ice abutting against the shore, but while we halted to cook breakfast, the wind opened a way for us. In the course of the morning we passed many heavy streams of ice, separated by lanes of open water, which would have afforded an easy passage for a ship. Having obtained a meridian observation for latitude, we re-embarked, and pulled for five miles through an open channel, to Point Cockburn, on the opposite side of a bay, which appeared to be four or five miles deep, and to be quite filled with drift-ice. Many deer were seen grazing near this point, but we did not stop to send a hunter in pursuit of them. We afterwards crossed several other indentations of the coast, skirted by reefs of limestone and low islets, and encamped on Chantry Island, lying close to the main shore, in latitude 68° 45′ N., longitude 114° 23′ W., having sailed thirty-nine miles in the course of the day. Two islands, lying opposite to our encampment, received the appellations of Manners-Sutton and Sir Robert Liston's Islands. The degree of motion in the ice, which was drifting between these islands and the shore, indicated a stronger current of both flood and ebb than we had hitherto seen.

Sunday, 6th. On the 6th, we commenced the day's voyage at three in the morning, but were compelled to put ashore soon afterwards by a stream of ice barring our way. At six o'clock, however, the flowing tide opened it sufficiently to enable us to push the boats along with poles, our progress being occasionally facilitated by the rocky reefs, which kept the heavier masses from pressing down upon us. Much of the ice lay aground, in nine fathoms, but none of it rose more than five or six feet above the surface of the water. We estimated the velocity of the flood tide, off some of the rocky points, at three miles an hour, and at such places we had much trouble in endeavouring to keep the boats clear of the drifting ice. The circular motion which the pieces occasionally acquired was particularly difficult to guard against, and had we not depended on the tongues of ice, which, lying deep under water, prevented the upper parts of the floes to which they belonged from coming in contact, we should scarcely have ventured amongst them. We did not, however, entirely escape, for the Dolphin was caught between a floe and a piece that lay aground, and fairly raised out of the water by the pressure, which broke one of her timbers and several of her planks. We put ashore on a small island to repair the damage, and during our stay Mr. Kendall had a meridian observation in latitude 68° 36½′ N. Another island, lying about two miles from the main land, was distinguished by the name of Aylmer Bourke Lambert, Esq., Vice-President of the Linnean Society. The sea water there was beautifully clear.

At half-past one, the Dolphin being again rendered sea-worthy, we prosecuted our voyage until five P.M., when the flood-tide set with such velocity round a rocky point, and brought so much ice with it, that we considered it prudent to put ashore. The violent eddies in the currents there, and the sudden approach and collision of the large masses of ice, reminded us forcibly of the poet's description of Scylla and Charybdis. The length of the day's voyage was twenty-one miles, and our encampment was situated in latitude 68° 32′ N., longitude 113° 53′ W. The temperature at nine P.M. was 60°.

Mr. Kendall and I took a walk of some miles along the shore, and were happy to observe the coast inclining to the southward, although no doubt now existed as to our accomplishing the voyage sufficiently early to allow us to cross the barren grounds, to the eastward of Great Bear Lake, before the cold weather set in. The flowering season for most of the plants on the coast was already past, but our route for the remainder of the distance to Bear Lake, inclining much to the southward, would naturally have the effect of prolonging to us the duration of the summer. A conspicuous hill, discovered in our walk, received the name of Mount Barrow, in honour of John Barrow, Esq., Secretary to the Admiralty; and two islands in the offing were named after Commanders Bayfield and Douglas, of the Royal Navy, to both of whom the officers of the Expedition were indebted for much assistance and personal kindness, in their progress through Canada. The interior of the country was flat, but the limestone formed cliffs on the shore two hundred feet high. From the form of the islands, I was led to believe that they consisted of trap rocks. Wollaston Land, as seen from the encampment, appeared to recede gradually from the main, and it sunk under the horizon, on a north-east bearing. By estimation, the most easterly part of it which we saw, is in latitude 68° 45′ N., and longitude 113° 53′. W. The navigation of the Dolphin and Union Straits would be dangerous to ships, from the many sunken rocks which we observed near the southern shore.

Monday, 7th. Embarking at two A.M. on the 7th, we crossed a deeply indented bay, which was named after Lieutenant-Colonel Pasley, of the Royal Engineers, to whose invention we owe the portable boat, named the Walnut-shell, which we carried out with us. On the east side of Pasley Cove there are some bold lime-stone cliffs, that form the extremity of a promontory, to which we gave the name of Cape Krusenstern, in honour of the distinguished Russian hydrographer. It lies in latitude 68° 23′ N., longitude 113° 45′ W., and is the most eastern part of the main land which we coasted. From a cliff, two hundred feet high, two miles to the southward of Cape Krusenstern, we had a distinct view of the high land about Inman's Harbour, on the western side of Cape Barrow, which was the most easterly land seen on this voyage, and lies in longitude 111° 20′ W. The space between Capes Barrow and Krusenstern is crowded with islands.

By entering George the Fourth's Coronation Gulf at Cape Krusenstern, we connected the discoveries of this voyage with those made by Captain Franklin on his former expedition, and had the honour of completing a portion of the north-west passage, for which the reward of five thousand pounds was established by his Majesty's Order in Council, but as it was not contemplated, in framing the Order, that the discovery should be made from west to east, and in vessels so small as the Dolphin and Union, we could not lay claim to the pecuniary reward.

While the party were at breakfast I visited Mount Barrow, which is a steep hill about three hundred feet high, surrounded by a moat fifty or sixty feet wide and twenty deep, and having a flat summit bounded by precipices of limestone. Three banks, like causeways, afforded the means of crossing the moat, and the hill altogether formed a remarkably complete natural fortification. The Esquimaux had marked most of the prominent points in this quarter, by erecting piles of stones similar to the cairns built for land-marks by the shepherds in Scotland. These erections were occasionally noticed, after doubling Cape Parry, but they were more numerous here. The ice which we saw this day was in form of loose streams, and offered no material impediment. Several wreaths of snow lay at the base of the cliffs that had a northern exposure, being the remains of that which had accumulated in the winter.

The latitude 68° 13′ N. was observed at noon on a low point which projected from some higher lands. From this point, which was named after Edward H. Locker, Esq., Secretary to the Royal Hospital at Greenwich, we had a view of Cape Hearne, the form of which I thought I recognised from my recollections of it on the former voyage. We reached Cape Hearne in the evening, having in the afternoon skirted a low and indented coast; a bay immediately to the north of it was named after Captain Basil Hall, of the Royal Navy. Cape Hearne itself is a low point, not visible from the mouth of the Coppermine; but the high land behind it, when seen from a distance, appears like a steep promontory, and is that designated as Cape Hearne in Captain Franklin's chart of his former voyage. The latitude of this cape is 68° 11′ N., and its longitude 114° 54′ W. The length of the day's voyage was forty miles. Many deer were seen here, and Ooligbuck killed a very fine one in the evening. After encamping I went a few miles into the interior, and found that the country was composed of limestone, which rose by a succession of terraces to the height of about three hundred feet above the sea. The heat of the day was considerable, the thermometer, when exposed to the rays of the sun, indicating 86°, without the bulb being blackened, or any other means used to retain the heat.

Tuesday, 8th.

Embarking early on the eighth, and passing through several loose streams of ice, some pieces of which were twenty-four feet thick, we landed at nine o'clock on a bold cape to prepare breakfast. It is formed of columnar greenstone, reposing on slaty limestone, and rising precipitously from the sea to the height of three hundred and fifty feet. I named this well marked point Cape Kendall, after my highly esteemed friend and companion, and had the pleasure of pointing out to him, from its summit, the gap in the hills at Bloody Fall, through which the Coppermine River flows. Mr. Kendall having taken the necessary bearings and sketches for the completion of his chart, we descended the hill to announce to the men, that a short traverse would bring us to the mouth of the Coppermine River. As we were aware of the disappointment which often springs from the premature excitement of hope, we had not previously acquainted them with our near approach to the termination of our voyage; fearing that an unfavourable trending of the coast, or an intervening body of ice, might protract it some days longer than we expected. The gratifying intelligence that we now conveyed to them, was, therefore, totally unexpected, and the pleasure they experienced found vent in heartfelt expressions of gratitude to the Divine Being, for his protection on the voyage. At noon the latitude of Cape Kendall was ascertained to be 67° 58′ N., and its longitude by reckoning was 115° 18′ W.

Re-embarking, we steered for the mouth of the Coppermine River with the sails set to a fine breeze, plying the oars at the same time, and on rounding Cape Kendall, we opened a magnificent inlet, or bay, rendered very picturesque by the manner in which its lofty cliffs came successively in sight as we crossed its mouth. We distinguished it by the name of our mutual friend and companion Captain Back. One of Couper's Islands, on which we landed, consists of greenstone, rising from the water like steps of a stair; and from its summit we perceived that a low piece of land, which, on the former voyage, had been mistaken for an island, was, in fact, the extremity of Point Mackenzie, and that Richardson River was merely a ravine, now dry.* Having reached the mouth of the Coppermine River, we encamped within a hundred yards of the position of the tents on Captain Franklin's former Expedition. Some half-burnt wood, the remains of the fires then made, were still lying on the spot; and I also recognised the Esquimaux stage, which we visited on that occasion, but there were no skins nor utensils on it now.

The completion of our sea voyage so early in the season was a subject of mutual congratulation to us all; and to Mr. Kendall and myself it was highly gratifying to behold our men still fresh and vigorous, and ready to commence the laborious march across the barren grounds, with the same spirit that they had shown in overcoming the obstacles which presented themselves to their progress by sea. We all felt that the comfort and ease with which the voyage had been performed, were greatly owing to the judicious and plentiful provision of stores and food which Captain Franklin had made for us; and gratitude for his care mingling with the pleasure excited by our success, and directing our thoughts more strongly to his party, the most ardent wishes were expressed that they might prove equally fortunate. The correctness of Mr. Kendall's reckoning was another source of pleasure. Having been de-

* Captain Franklin has since transferred the name of Richardson to the Bay between Point Mackenzie and the mouth of the Coppermine River.

prived of the aid of chronometers, by the breaking of the two intended for the eastern detachment of the Expedition, during the intense winter cold, our only resource for correcting the dead reckoning was lunar observations, made as frequently as opportunities offered; yet when we approached the Coppermine River, Mr. Kendall's reckoning differed from the position of that place, ascertained on Captain Franklin's former Expedition, only twenty seconds of time, or about two miles and a half of distance, which is a very trifling difference when the length of the voyage and the other circumstances are taken into consideration. The distance between Point Separation and the mouth of the Coppermine River, by the route we pursued, is nine hundred and two geographical miles.

In our progress along the coast no opportunity was omitted of noting the times of high-water, and a tide-table drawn up by Mr. Kendall, is given in pages 236, 237. We nowhere observed the rise of the tide to exceed twenty-two inches, and in some places it was not more than eight or nine; but the velocity of the flood and ebb was greater than could have been expected from so small a rise. Off the Alluvial Islands, lying between the outlets of the Mackenzie River and Esquimaux Lake, it was in the strength of the flood about a mile an hour; at Cape Bathurst it exceeded a mile and a half; and in the Dolphin and Union Straits it was fully three miles. The stream of the flood set every where from the eastward.

The variation of the magnetic needle, which was forty-six degrees easterly at Point Separation, attained to 50° at Refuge Cove, 53° at Point Maitland, and 56° at Cape Parry; after which it gradually decreased as we went to the south-east; and at the mouth of the Coppermine, it was 48°.

We saw no ice that would have much impeded a ship, except between Sir George Clerk's Island and Cape Bexley, where it was heavy and closely packed. The appearance, however, of lanes of open water towards Wollaston Land, opposite to Cape Bexley, induced us to think that there might be a good passage for a ship on the outside of the ice, which lined the south shore, and which seems to have been packed into the indentations of the coast by the strong north-west winds that had prevailed for some days. A ship would find shelter amongst the islands of George the Fourth's Coronation Gulf, in Back's Inlet, in Darnley Bay, and amongst Booth's Islands, lying off Cape Parry; but the bottom, at the latter place, is rocky, and there are many sunken rocks along the whole of that coast. To the westward of Cape Parry, we saw no ship harbours, and the many sand-banks skirting the outlets of Esquimaux Lake would render it dangerous for a ship to approach the shore in that quarter. There is such an abundance of drift-timber on almost every part of the coast, that a sufficient supply of fuel for a ship might easily be collected, and wherever we landed on the main shore we found streams or small lakes of fresh water. Should the course of events ever introduce a steam-vessel into those seas, it may be important to know that in coasting the shores between Cape Bathurst and the Mackenzie, fire-wood sufficient for her daily consumption may be gathered, and that near the Babbage River, to the westward of the Mackenzie, a tertiary pitch-coal exists of excellent quality, which Captain Franklin describes as forming extensive beds.

The height to which the drift-timber is thrown up on the shores at the western entrance of the Dolphin and Union Straits is, I think, an indication of an occasional great rise in the sea, which, as the tides are in comparison so insignificant, I can ascribe only to the north-west winds driving the waters of an open sea towards the funnel-shaped entrance of the straits. If this view is correct, Wollaston Land probably extends far to the north, and closely adjoins to Banks' Land, or is connected with it. Captain Parry found the strait between Melville Island and Banks' Land obstructed by ice, and this will naturally be generally the case, both there and in the Dolphin and Union Straits, if they form the principal openings through a range of extensive islands, which run north and south, and bound a large tract of sea, comparatively free from land. The heat of the summer in that quarter seems to be always or almost always sufficient to admit of the ice breaking up, but not powerful enough to dissolve it entirely. Hence the loose ice driven about by the winds, and carried to the lee-side of the wider expanses of sea, is firmly packed in the narrow straits and winding passages amongst the islands, from whence it can be dislodged only by a concurrence of very favourable circumstances, and where the waste by the solar rays is replaced by every breeze blowing from the open sea. The north-west winds being the strongest and most prevalent in the latter part of the summer, it is at the western end of a strait that the ice is most frequently and closely packed. Captain Parry remarks that "there was something peculiar about the south-west extremity of Melville Island, which made the icy sea there extremely unfavourable to navigation, and which seemed to bid defiance to all efforts to proceed farther to the westward in that parallel of latitude." The Dolphin and Union Straits hold out greater prospects of success for a similar attempt, not only from their more southern position, but from the strong current of flood and ebb which flows through them and keeps the ice in motion.

We noticed on the coast about one hundred and seventy *phænogamous*, or flowering plants, being one-fifth of the number of species which exist fifteen degrees of latitude farther to the southward. The grasses, bents, and rushes, constitute only one-fifth of the number of species on the coast, but the two former tribes actually cover more ground than all the rest of the vegetation. The cruciferous, or cress-like tribes afford one-seventh of the species, and the compound flowers are nearly as numerous. The *shrubby plants* that reach the sea-coast are the common juniper, two species of willow, the dwarf birch (*betula glandulosa*), the common alder, the hippophae, a gooseberry, the red bearberry (*arbutus uva ursi*), the Labrador tea plant, (*ledum palustre*,) the Lapland rose (*rhododendron lapponicum*,) the bog whortleberry (*vaccinium uliginosum*,) and the crow-berry (*empetrum nigrum*.) The kidney-leaved oxyria grows in great luxuriance there, and occasionally furnished us with an agreeable addition to our meals, as it resembles the garden sorrel in flavour, but is more juicy and tender. It is eaten by the natives, and must, as well as many of the cress-like plants, prove an excellent corrective of the gross, oily, rancid, and frequently putrid meat, on which they subsist. The small bulbs of the Alpine bistort (*polygonum viviparum*,) and the long, succulent, and sweet roots of many of the *astragaleæ*, which grow on the sandy shores, are eatable; but we did not learn that the Esquimaux were acquainted with their use. A few clumps of white spruce-fir, with some straggling black spruces and canoe birches, grow at the distance of twenty or thirty miles from the sea, in sheltered situations, on the banks of rivers.

ABSTRACT of the Meteorological Register, kept by the Eastern Detachment, in their Voyage between the Mouths of the Mackenzie and Coppermine Rivers.

Date.	Temperature in the Shade.			Direction of the Winds.	Weather and Remarks.
	Lowest.	Highest.	Mean.		
July 9	32	38	35	East; NEbE.	Fresh breezes. Clear sky with fog over the ice.
10	45	57	51	ESE.	Strong breezes, clear weather.
11	42	51	43	East.	Strong breezes, clear sky, and bright sun.
12	45	50	47	Do.	Ditto, ditto.
13	46	57	52	East; SE.	Moderate breezes. Clear sky; rain in the night.
14	42	42	42	West.	Heavy gales. Thick fog.
15	52	57	55	Nearly calm.	Very fine weather.
16	38	55	47	South.	Moderate breezes. Cloudy A.M., clear P.M.
17	50	62	54	West; North.	Fog A.M. When wind veered to north cleared up. Temperature of sea 55°.
18	45	56	50	South; East.	Light airs A.M.; fresh breezes P.M.; calm in the night.
19	44	54	49	East; West.	Fresh breezes and cloudy A.M. Four P. M. West wind and foggy weather.
20	46	50	48	NW.; WNW.	Foggy; fresh breezes A.M. Increased to a strong gale P.M.
21	42	48	46	WNW.; NW.	Fresh breezes and foggy A.M. Fine and clear P.M.
22	45	47	46	South.	Fresh breezes A.M. Fine weather P.M.
23	46	58	52	SW.	Moderate and cloudy. Many Musquitoes.

Meteorological Register, &c.—Concluded.

Date.	Temperature in the Shade.			Direction of the Winds.	Weather and Remarks.
	Lowest.	Highest.	Mean.		
July 24	50	76	66	West.	Moderate breezes. Foggy A.M. Occasionally hazy P.M. Myriads of Musquitoes.
25	45	66	55	South; NE.	Fine A.M. Strong gales and partial fogs P.M.
26	35	47	41	NE.	Strong gales and clear. Temperature of sea 35°.
27	35	45	40	ENE.	Moderate.
28	35	50	42	Calm; North.	Fine clear weather.
29	37	41	38	WNW.	Moderate breezes; foggy.
30	36	40	38	WNW.	Fog hanging over the ice; clear inland; moderate breezes.
31	38	45	41	NW.	Moderate breezes; occasionally hazy, fog over the ice.
Aug. 1	43	50	48	West.	Moderate breezes; hazy to seaward.
2	34	50	41	West; variable.	Hazy and occasionally foggy.
3	38	43	40	NE.; East.	Light breezes and clear.
4	38	55	46	NE.; ESE.	Fine clear weather.
5	39	56	47	EbS.; WSW.	Do. Moderate P.M.
6	42	56	47	South; variable.	Do.
7	36	68	52	SSE.	Fine and very clear. Temperature in the sun 86°.
8	44	60	52	North.	Do.
	41.45	51.92	46.48		

PROCEEDINGS OF THE EASTERN DETACHMENT CONCLUDED.

CHAPTER IV.

Ascend the Coppermine River—Abandon the Boats and Stores—Commence the Land Journey—Cross the Copper Mountains and Height of Land—Meet Indians who bring Provisions—Arrive at Great Bear Lake—Detained by want of a Boat—Send out Hunters—Arrival of Beaulieu—Collect the Party, and proceed to Fort Franklin—Conclusion.

Wednesday, 9th.

At four o'clock in the morning of the 9th of August, we left our encampment at the mouth of the Coppermine River, and proceeded in the boats to Bloody Fall, a distance of about eleven miles. The river was very low, and, in many places, there was scarcely water enough for our boats, which did not draw more than fourteen inches. On the preceding evening an Esquimaux dog had come to our encampment: his meagre aspect showed that he had fared badly, and hunger had rendered him so tame that he readily ate from our hands. After following us a considerable way up the river he left us; and we found, on our arrival at Bloody Fall, that a party of Esquimaux had just quitted that place; probably having discovered us from a distance.

The Coppermine River, for forty miles above Bloody Fall, flows over an uneven stony bed, betwixt precipitous rocky walls, and is full of rapids. It is totally impracticable to ascend it in boats having a greater draught of water than a few inches; and even a small canoe must be frequently carried over land for considerable distances, to avoid the numerous obstacles which occur. It was necessary, therefore, that we should leave at this place the Dolphin and Union, and every thing that was not absolutely necessary for our journey. We determined, however, on taking with us Colonel Pasley's canvass boat, the Walnut-shell, in the hope of its occasionally relieving the men of their burdens for a short time, should any part of the river admit of its use. The afternoon was employed in arranging the loads for crossing the barren grounds. Twenty pounds of pemmican were allotted to each man, and the packages of maccaroni, arrow-root, portable-soup, chocolate, sugar, and tea, were equally distributed; together with the nautical almanack, astronomical tables, charts, two fishing nets, the collection of plants, specimens of rocks, and the portable boat, kettles, and hatchets; all of which, with the blankets, spare shoes, guns, and ammunition, made a load of about seventy-two pounds a man. Mr. Kendall undertook to carry the sextant and azimuth-compass; and I took the artificial horizon and a package of paper for drying plants, besides which we each carried a blanket, gun, and ammunition. As I feared that some of the party would over-rate their strength, and, through a desire of saving some favourite article, load themselves too heavily at the outset, which could not fail to prove very injurious to the regularity and speed of our march, I informed them, that, as soon as we were at a convenient distance from our present encampment, I should halt and examine all their bundles.

The boats were drawn up on shore, out of the reach of any flood, and the remainder of the articles, that we had brought to give the Esquimaux, were put into boxes and placed in the tents, that they might be readily found by the first party of that nation that passed this way. They consisted of fish-hooks, lines, hatchets, knives, files, fire-steels, kettles, combs, awls, needles, thread, blue and red cloth, gartering, and beads, sufficient to serve a considerable number of Esquimaux for several years. The tents were securely pitched, and the Union Jack hoisted, partly for the purpose of attracting the attention of the natives, and partly to show them the mode of using the tents, which may prove to be very useful in their summer journeys. That no accident might occur from the natives finding any of our powder, all that we did not require to take with us was thrown into the river.

Thursday, 10th.

At six o'clock on the morning of the 10th, after the men had been down to the beach to take a last look of our little boats, we began our march to Bear Lake, intending to keep on the banks of the Coppermine as far as its bend at the Copper Mountains, and to strike from thence straight across the hills for the mouth of Dease's River, which falls into the north-east arm of Bear Lake. We set off at a pretty quick pace, and the first hill, after leaving our encampment, being steep, tried the wind of most of the party, so that the few who had loaded themselves with superfluous articles, were glad to throw them away during a short halt on its summit, and when I examined their packages, at the next resting-place, I found little to reject. A path beaten by the rein-deer and the Esquimaux conducted us down the southern face of this range of hill to the plain beneath, when we halted to prepare breakfast, and to make some further arrangements, as several of the party, being unaccustomed to carry loads, advanced slowly. After breakfast the portable boat was put together, and the baggage being placed in it, we endeavoured to tow it up the river, but found this to be impracticable, owing to the badness of the towing-path, the numerous high cliffs which bound the stream, and the form of the boat, which permitted the water in strong rapids to flow over its bows. This boat was admirably adapted for the purpose for which it was constructed by Colonel Pasley, that of crossing a river or lake, as we had ascertained by previous trials; but we knew that no river, except such as we could ford, could occur on our route to Bear Lake; and I, therefore, determined on leaving it, together with half a bag of arrow-root, and five muskets, by which the loads were reduced about fifteen pounds a man. The march was then resumed with alacrity, and, notwithstanding that the day was hot and sultry, we proceeded with greater speed and satisfaction. Mr. Kendall walked at the head of the line at a steady pace, halting for five minutes every half hour to rest the party, and prevent straggling. At five we encamped, having marched about six miles in a direct line. The route throughout the journey was regulated, from time to time, by our taking the bearing of a distant hill, or other conspicuous object, by the compass, and walking directly for it; and the distance was estimated by noting the time and guessing the rate of our march. Of this, which was in general a little more than two miles an hour, previous practice had enabled us to judge so correctly, that the estimate seldom erred more than a mile a day. The error, whatever it was, was always corrected at noon, when the latitude was observed, and the course and distance were then calculated anew.

During the day several small herds of rein-deer were seen, but I would not permit any one to leave the line of march to go in pursuit of them; after encamping, however, M'Leay killed a fine buck. A solitary stunted spruce-fir grew near our encampment, and the most northerly clump on the river was seen about two miles to the southward. When supper was over and a watch set, we stretched ourselves on the ground, and soon sunk into sound sleep. The temperature at sunset was 62°.

Setting out on the 11th, at six in the morning, we halted to breakfast at nine, and Mr. Kendall took an observation at noon, in latitude 67° 33′ N. We encamped at half past five P.M. amongst some small pines. The day was fine, and a fresh easterly wind rendered it agreeable for walking; but the men were much annoyed by their burdens, and appeared jaded when we halted for the night. Their loads could not have exceeded fifty-two pounds each, but the frequent ascent and descent in crossing the small hills that lay in our way, and the occasional sponginess of the ground, and insecurity of footing, rendered marching much more laborious than it would have been on a hard English highway. The direct distance travelled this day was about twelve miles. We saw many gray Arctic marmots (*Arctomys Parryi*,) sporting near their burrows, and a little terrier dog, which had been our fellow voyager from England, showed much dexterity in cutting off their retreat, and succeeded in catching several of of them. The dog's long confinement in the boat rendered the exercise he now took very fatiguing, and when we halted for the night he was the most tired of the party. Many young rein-deer were also seen, and after we encamped Ooligbuck killed one. The temperature in the evening was 50°, but the night was cold.

Friday, 11th.

Our march on the 12th was rendered pleasant by a cool northerly breeze, and the men being now familiar with their loads, which had also suffered some dimunition by the preceding evening's repast, we made a more rapid progress. The length of the march was seventeen miles, being, exclusive of the half hourly halts and the time occupied by breakfast, at the rate of two miles and a half an hour. In the course of the day we crossed several ridges of the Copper Mountains to avoid a bend of the river. The Whisky-John (*corvus Canadensis*) visited our encampment in the evening for the first time since we left the Mackenzie.

Saturday, 12th.

On the 13th, commencing the day's march at five A.M., we walked along the banks of the river until nine, when we halted to prepare breakfast, at the place where Captain Franklin encamped on the 11th of July 1821. After breakfast we forded the small stream, on the banks of which several pieces of native copper and some copper ore were found on the former Expedition. A quantity of ice formed by snow, consolidated by the oozing of the stream, still remained in the bed of this rivulet.

Sunday, 13th.

At noon the latitude was observed in 67° 13′ N., and as we were now on the spot where the Coppermine makes the nearest approach to the north-east arm of Bear Lake, we decided on striking directly from this place to the mouth of Dease's River, and the course and distance were accordingly calculated. Our route lay over rocks of old red sandstone, clay-slate, and greenstone disposed in ridges, which had a direction from E.S.E. to W.N.W. The sides of many of the ridges were precipitous, and their uneven and stony summits were two hundred or two hundred and fifty feet high. The valleys were generally swampy and abounded in small lakes. A few scattered and thin clumps of pines existed in the more sheltered spots, but the country was, in general, naked. Several burrows of wolves were seen in the mountains. We crossed two small streams in the course of the day, flowing towards the Coppermine, and encamped at four P.M. on the banks of a small lake. Sand-flies, the first we had seen this season, were numerous and troublesome in the evening, the temperature then being 53°.

Monday, 14th. Setting out at five A.M. on the 14th, we halted to breakfast at nine, after a pretty brisk walk through a country entirely destitute of wood. Some partridges, which were so tame as to be easily killed with stones, furnished us with an agreeable variety of diet. A meridional observation was obtained in latitude 67° 10′ N. In endeavouring to get round the south end of a small chain of lakes, which lay in our route, we were stopped by a narrow stream about six feet deep, flowing from them towards the Coppermine River; but, on sounding the lake a little way from the head of the stream, we found that it was fordable without difficulty. We marched to a late hour in search of fuel to cook some deer's meat, which M'Leay had procured in the course of the day, and were fortunate in at length finding a wooded valley on the banks of a small stream, that fell into the chain of lakes which we had crossed. It is probably this river, and chain of lakes, that the Indians ascend from the Coppermine River in canoes to the height of land which they cross on their route to Bear Lake,* The ridges of hill over which we marched on this day consisted of spotted sandstone and porphyry. The temperature in the evening was 47°, and the night was frosty. Two white wolves took a survey of our bivouack, but did not venture within gun-shot.

Tuesday, 15th. Starting on the 15th at five o'clock, we marched until eight, when we halted to breakfast. The air felt very cold, although the thermometer was

* Franklin's First Journey to the Polar Sea, p. 337.

not below 39°. In the early part of the day we crossed some ridges of sandstone, and towards noon we travelled over granite, similar to that which abounds in the neighbourhood of Fort Enterprize. Much wood was seen in a valley far to the westward, but the hills over which our course lay were quite naked. The bog whortleberry (*vaccinium uliginosum,*) however, grew abundantly on these hills, and as its fruit was now in the highest perfection, the men at every resting-place threw themselves down, and indulged freely, without sustaining any injury.

In the afternoon our route was over nearly horizontal strata of spotted sandstone and conglomerate. About three o'clock we had gained the summit of the height of land separating the Coppermine River from Great Bear Lake, and obtained from it an extensive view of a lower and well wooded country; but all the grounds in our immediate neighbourhood consisted of barren sandstone strata. After looking in vain for a comfortable sleeping-place, as the night threatened to be stormy, and a moist and cold fog was setting in, we were obliged to content ourselves with building a rude shelter with blocks of sandstone; and to use for firing a black lichen (*cornicularia divergens,*) which, fortunately, grew plentifully in the crevices of the rock. The distance walked this day was about fourteen miles. We had no meridional observations, because the sky was obscured.

We had supped, and most of the men had retired to rest, when Mr. Kendall, in sweeping the horizon with his telescope, saw three Indians coming down a hill, and directing their steps towards us. More moss was immediately thrown on the fire, and the St. George's ensign hoisted on the end of a musquet, to point out to the comers who we were; but as they hid the youngest of their number in a ravine, at the foot of the hill, and the two seniors seemed to approach slowly and with suspicion, Mr. Kendall and I went unarmed to meet them. They came up, one with his bow and arrows in his hand, and the other with his gun cocked; but as soon as they recognised our dress, which was the same that I had worn in our voyage round Bear Lake, the preceding autumn, when I had seen most of the Hare Indian tribe, they shouted in an ecstacy of joy, shook hands most cordially with us, and called loudly for the young lad to come up. The meeting was no less gratifying to us: these people had brought furs and provisions to Fort Franklin in the winter, and they now seemed to be friends come to rejoice with us on the termination of our voyage. We learned from them, partly by signs, and partly from the little we understood of their language, that by the advice of It-chin-nah, the Hare Indian Chief, they had been hunting for some time in this neighbourhood, in the hopes of falling in with us on our way from the sea; that they would give us all the provision they had collected, accompany us to Bear Lake, and warn all the Indians in the neighbourhood of our arrival. They appeared much surprised, when, placing the compass on the ground, we showed them the exact bearing of the mouth of Dease's River; and they were not able to comprehend how we knew the way in a quarter through which we had never travelled. They said, however, that they would conduct us in the morning to the Indian portage road, where we would have better walking than by keeping the direct route across the hills. We had reserved but little that we could present to these kind people, though every one contrived to muster some small article for them, which they gratefully received. They were dressed, after the manner of their tribe, with fillets of deer-skin round their heads and wrists, and carried in their hands a pair of deer's horns and a few willow twigs, which are all serviceable in enabling them to approach the rein-deer, in the way described by Mr. Wentzel in the Narrative of Captain Franklin's former voyage.

Ooligbuck, who had gone out to hunt, returned in the night. He met an Indian who had just killed a deer with an arrow, and had tried to persuade him to come to us; but neither of them understood the other's language, and the Indian, probably terrified by the sight of an Esquimaux armed with a gun, presented him with a piece of the deer's meat, and then made off in an opposite direction. Many of the Hare Indians abstain from visiting the forts for several years, and it is possible that this one had not heard of us, or at least had not received a distinct account of our intention of returning this way, and of our having an Esquimaux with us. Our Indian friends told us that they did not know that any of their countrymen were hunting in the direction which Ooligbuck pointed out.

Wednesday, 16th. On the 16th a thick fog prevented us from quitting our bivouack until seven o'clock, when the Indians led us down the hill about a mile to the portage road, and we resumed the precise line of march that we had followed from the Coppermine River, (S. 63° W.) Such of our Highlandmen as had been in the service of the Hudson's Bay Company, and, consequently, knew from experience the difficulty of travelling through a country without guides, could not help expressing their surprise at the justness of the course we had followed. We had not concealed from them, that from want of observations, or from the difficulty of estimating the distance walked, we might err a mile or two in our reckoning, so that they were prepared, on our reaching Bear Lake, to turn a little to the right or left in search of the river; but they had scarcely hoped to have reached that point without having to perform a single mile of unnecessary walking.

The portage-road conducted us in a short time to the principal branch of Dease's River, on the banks of which, at the distance of six miles from our encampment, we halted to breakfast. The stream there receives another branch, but it is fordable without difficulty, being nowhere much above knee-deep. A little way further to the westward, however, it is less rapid, and forms frequent lake-like expansions. Our march from last night's encampment was over sandstone rocks, and down a pretty rapid ascent. The ground was barren in the extreme, except at our breakfasting place, where there was a convenient clump of wood and a profusion of whortleberries. Having finished this meal, we resumed the march, with the intention of halting a few miles further on, that our Indian friends might rejoin us with their provision, which lay in store to the southward of our route. We therefore encamped at half past two o'clock in a pleasant pine clump, and immediately set fire to a tree to apprize the Indians of our situation. They arrived at sunset, heavily laden with tongues, fat, and half-dried meat; and M'Leay also killed two deer after we encamped, so that we revelled in abundance. The length of the day's journey was fourteen miles, and the estimated distance of the mouth of Dease's River twenty miles.

Thursday, 17th.

The provisions obtained from the Indians being distributed amongst the men, we commenced the march at five o'clock in the morning, and walked, until the usual breakfasting hour, over a piece of fine level ground. A range of sandstone hills rose on our left, and the river ran nearly parallel to our course on the right, but we walked at the distance of one or two miles from it, to avoid its windings and the swampy grounds on its borders. Pine-trees grow only in small detached clumps on its south bank; but the uneven valley, which we saw spreading for ten or twelve miles to the northward, was well wooded. The Needagazza Hills, which lie on the north shore of the Bear Lake, closed the view to the westward. Several columns of smoke wese seen to the westward, and one to the southward; the latter, the Indians informed us, was made by It-chinnah. We breakfasted on the banks of a small stream, where the whortleberry bushes were loaded with fruit of a finer flavour than any we had previously met with. At noon we crossed a hill, on the summit of which Mr. Kendall had an observation, that placed it in 68° 58′ of north latitude. Our route afterwards led us across several deep ravines close to the river, which there runs by the base of some lofty cliffs, of light red sandstone, and we pushed on in great spirits, and at a rapid pace, with the intention of reaching Bear Lake that evening; but the Indians complaining that they were unable to keep up with us, we halted at three P.M. Several trees were then set on fire to apprize It-chinnah and his party of our approach; and, after supper, I went to the summit of a hill, and readily recognised the islands in Dease's Bay of Bear Lake, from their peculiar form and disposition.

Friday, 18th. Setting out at three A.M. on the 18th, the Indians conducted us over a rising ground, covered with white spruces, to a bay of the Great Bear Lake, about a mile from Dease's River. After breakfast, our stock of provisions being examined, it was found that we had two days' allowance remaining. A party was next sent to Dease's River to make a raft for setting the two nets, and they were also directed to look for traces of Beaulieu and his party. He had been ordered by Captain Franklin to leave the fort on the 6th of August, and to make the best of his way to the rendezvous, where he was to remain to the 20th of September. The length of his voyage, allowing for two or three days detention by adverse winds, was not expected to exceed seven or eight days, nor to be protracted, under any circumstances, beyond ten or twelve. We had, therefore, reason to suppose that he might have reached Dease's River by this time. He was fully aware of the inconvenience that we might experience, should we reach the appointed spot and find no provisions there; and to stimulate him to make as much haste as possible, I had promised him a fowling-piece, on condition that we found him waiting for us on our arrival. Huts were made to sleep in, and several trees set on fire to point out our position to the Indians in the neighbourhood.

Saturday, 19th. The mossy ground near our encampment caught fire in the night, and the flames spread so rapidly that we were obliged on the morning of the 19th, to move to the banks of the river, where we made new huts. Owing to the loss of a hatchet in driving the stakes, only one net had been set the preceding evening, and in it we took eight carp. The raft being made of green wood was not sufficiently buoyant, and a new one was, therefore, constructed this day of dried timber. The carp afforded a breakfast for the party, and supper consumed all our deer's meat, together with a portion of the remainder of the pemmican. The young Indian went off in the afternoon in quest of It-chinnah's party. A strong easterly wind blowing all this day, was adverse to Beaulieu's advance.

On Sunday, the 20th, prayers were read, and thanks returned to the Almighty for his gracious protection and the success which had attended our voyage. The nets yielding seventeen pike, carp, and white fish, provided an ample breakfast for the party, and before supper time the young Indian returned with two of his countrymen, bringing meat sufficient for three days consumption. Part of it was the flesh of the musk-ox, which was fat and juicy, but had a high musky flavour. We had seen none of these animals on our march from the Coppermine River, although we frequently noticed their foot marks. Frequent squalls during the day brought much rain, but the huts which we had made of pine branches kept us dry. We could not but consider ourselves fortunate in having had no rain in the journey overland, when there was not sufficient wood to afford us the shelter we now experienced.

Sunday, 20th.

On the 21st the nets yielded sixteen fish, which were enough for breakfast. Mr. Kendall crossed the river on a raft, and went to the top of a hill to the westward to look for Beaulieu; and, by way of keeping the men employed, I sent M'Leay and some of our best hunters in quest of deer, and set the carpenter and the remainder of the party to make oars. Our Indian friends left us to warn some more of their countrymen, of our situation, and five others arrived in the evening, bringing meat and large basketfuls of whortleberries. M'Leay and the other hunters returned without having seen any deer.

Monday, 21st.

To secure a stock of provision for our journey to the fort, in the event of any accident preventing the arrival of the boat, I resolved to send half the party on a distant excursion, and on the 22nd, Gillet, M'Leay, M'Duffie, M'Lellan, and Ooligbuck, were despatched to hunt in the neighbourhood of Limestone Point, on the north shore of the lake, with orders not to extend their excursions beyond Haldanes River, which falls into the lake about sixty miles to the westward of Dease River. If they went on to Haldanes River, they were to set up a mark on Limestone Point, that I might know whether they had passed or not. They took with them a small supply of provision, and an Indian guide. In the evening two Indians came with more meat. They were desirous of being paid with ammunition, which they much need-

Tuesday, 22nd.

ed, but we had none to give them, and they cheerfully took our notes of hand for payment, on their arrival at the fort in the winter.

Wednesday, 23rd. The 23d day of August having passed away like the four preceding ones, in anxious expectation of Beaulieu's arrival, I began to apprehend that some serious accident had happened to his boat, and to fear that we should be obliged to walk round the Lake to the Fort. The distance exceeding three hundred miles, we could not expect to accomplish it in less than three weeks, and not without much fatigue and suffering, for the men's stock of shoes was nearly exhausted, their clothing ill adapted for the frosty nights that occur in September, and deer do not frequent, at this season, much of the country through which our route lay. I naturally looked forward to such a march with uneasiness, yet, as the season was drawing to a close, I determined not to delay setting out beyond the 28th, when I intended to engage some Indians as guides, and to take with us as much dried meat as we could carry. The wind blew from the south-west this day, and we were much tormented by sand-flies.

Thursday 24th. On the evening of the 24th, as we were about to retire to bed, having given up all hopes of Beaulieu's arrival that day, we heard people talking in the direction of the mouth of the river, and soon afterwards saw a boat and several canoes. A musket being fired to show them our position, they steered for the encampment, and landed opposite to the huts. They proved to be Beaulieu's party, consisting of four Canadians, four Chipewyan hunters, and ten Dog-ribs, which, with their wives and children, amounted to about thirty in all. We learnt from Beaulieu, that he had been sent off from the Fort by Mr. Dease, on the 6th, with strict injunctions to proceed to the rendezvous with his utmost speed; but he pleaded the badness of the weather and the adverse winds as the cause of his delay. He had not seen the five men I sent off on the 22d, though he had noticed a fire in a bay near Limestone Point, which I had no doubt was made by them; I I therefore embarked directly to rejoin them at that place, accompanied by Mr. Kendall and the remainder of our party, two of the Canadians, and an Indian named the Babillard; directing Beaulieu to stay at the huts until he heard from us again. We rowed all night, and soon after day-break reached the spot where the fire had been made, but found no marks to indicate which way our men had gone: neither was there any mark at Limestone Point; I therefore caused a large fire to be made at the latter place, and remained there the whole day.

Our people not appearing on the 26th, I returned in the boat to Dease River, leaving Mr. Kendall and the Babillard at Limestone Point. Beaulieu had seen nothing of the absentees, and it was therefore evident that they had gone on to Haldane River, whither I resolved to proceed in search of them; but that they might not suffer from want of food, if by any chance we missed them, I directed Beaulieu's party to remain where they were, until I sent them permission to depart by two Canadians, whom I took with me on purpose in a small canoe. Mr. Dease had directed Beaulieu to go to M'Tavish Bay to hunt deer, and dry meat for the fort, as soon as we arrived; and as the boat was well adapted for carrying dried provision, I now exchanged it with his north canoe.

Saturday, 26th.

We rejoined Mr. Kendall at Limestone Point at day-break on the morning of the 27th, and afterwards paddled along the coast until two P.M., when a strong head-wind obliged us to put ashore. As soon as we landed, I set out with the Babillard for Haldane River, carrying a small quantity of pemmican, lest the people should be in want of food; and after a walk, or rather a run, of five miles, I had the happiness of finding them all well, and with plenty of provisions, as they had killed six deer. Their Indian guide had taken them a little inland, by which they had missed Limestone Point; but they were very sorry it had so happened, when they learned the anxiety they had occasioned to Mr. Kendall and myself, by their not erecting the mark there as they had been directed to do. The wind moderating after sunset, Mr. Kendall joined us with the two canoes, so that the party was again happily reunited. On Monday the 28th, I sent back the small canoe with the Babillard and two Canadians, to join Beaulieu, and proceed with the rest of the party in the larger canoe to Fort Franklin, where we arrived on Friday, the 1st of September, and received a warm welcome from Mr. Dease, after an absence of seventy-one days, during which period we had travelled by land and water one thousand seven hundred and nine geographical, or nineteen hundred and eighty statute miles.

Sunday, 27th.

Monday, 28th.

Having now brought the Narrative of the proceedings of the Eastern Detachment to a conclusion, the pleasing duty remains of expressing my gratitude to the party for their cheerful and obedient conduct. Not a murmur of discontent was heard throughout the voyage, but every individual engaged with alacrity in the laborious tasks he was called upon to perform. Where all behaved with the greatest zeal, it would be invidious to particularize any; and I am happy in having it in my power to add, that since our return to England, Gillet, Fuller and Tysoe, who were in His Majesty's service previous to their being employed on the Expedition, have been rewarded by promotion. Our good-natured and faithful Esquimaux friend Ooligbuck, carried with him to his native lands the warmest wishes and esteem of the whole party. His attachment to us was never doubtful, even when we were surrounded by a tribe of his own nation.

The general abilities and professional skill of my companion, Lieutenant Kendall, are duly appreciated in higher quarters, and can derive little lustre from any eulogium from me; but I cannot deny myself the gratification of recording my deep sense of the good fortune and happiness I experienced in being associated with a gentleman of such pleasing manners, and one upon whose friendly support and sound judgment I could with confidence rely, on occasions of difficulty and doubt inseparable from such a voyage.

End of Dr. Richardson's Narrative of the Proceedings of the Eastern Detachment.

TABLE of the Distances travelled by both Branches of the Expedition, and of the extent of their Discoveries in 1827.

BY THE WESTERN PARTY.	*Statute Miles.*
From Fort Franklin, by Fort Norman, to Point Separation (river course)	525
Point Separation to Pillage Point, at the Mouth of the Mackenzie	129
Pillage Point to Return Reef (sea-voyage out)	374
Return Reef, back to Fort Franklin, including Peel River	1020
Distance travelled by the Western Party in July, August, September, 1826.	2048

BY THE EASTERN PARTY.	
From Fort Franklin to Point Separation, along with the western party	525
Point Separation to Point Encounter (river course.)	159
Encounter to the Coppermine River (sea-voyage*)	863
The mouth of the Coppermine, over land to Fort Franklin	433
Distance travelled by the Eastern Party in July and August, 1826	1980

EXTENT OF THE DISCOVERIES OF THE WESTERN PARTY IN 1826.	
From Point Separation to the mouth of the Mackenzie, by a western branch, not previously known	129
Pillage Point by the sea-coast to Point Beechey, which was seen from Return Reef	391
Peel River and a branch of the Mackenzie surveyed for the first time, on the return	90
	610

EXTENT OF THE DISCOVERIES OF THE EASTERN PARTY IN 1826.	
From Sacred Island to Point Encounter, being a portion of the river lying to the eastward of Mackenzie's route	37
Point Encounter, along the coast to the Coppermine River	863
The Copper Mountains, overland to Bear Lake	115
	1015

* All the distances mentioned in the narrative of the proceedings of the eastern detachment, are geographical miles.

TABLE of Times of High Water, reduced to Full and Change, by E. N. Kendall, Lieutenant, R. N.

Date.	Names of Places of Observations.	Geographic Position.		Times of High Water.		Winds.		General Remarks.
		Lat. N.	Lon. W.	Observed.	Reduced to full & change.	Direction.	Force.	
		° ′	° ′	h. m.	h. m.			
1825. Aug. 16	Garry Island - -	69 29	135 41	1 0 P.M.	10 19	N.E.	6	No ice visible.
1826. July 9	Point Toker - - -	69 38	132 18	4 25 P.M.	1 45	N.E.b.E.	5	Loose ice covering the sea. Rise of water 20 in.
10	Bay between Points Toker and Warren	69 43	131 58	5 0 P.M.	0 56	East.	8	Heavy pieces of ice.
12				6 48 P.M.	1 48	- -	5	Little ice visible.
13	Atkinson Island - -	69 55	130 43	7 0 P.M.	0 32	S.E.	1	Rise and fall 18 inches.
14	Browell Cove - -	70 00	130 20	7 0 P.M.	1 12	West.	6	Very little rise and fall.
18	Point Sir P. Maitland	70 08	127 45	3 15 A.M.	3 47	Calm.	- -	In the mouth of Harrowby Bay, round which the tide appeared to flow.
19	Near Cape Bathurst	70 33	127 21	1 30 A.M.	1 28	E.S.E.	6	Flood setting from the Eastward. Rise and fall 14½ inches.
20	Point Fitton - - -	70 11	126 14	4 00 A.M.	3 18	N.W.	6	Rise and fall 13 inches.

TABLE of Times of High Water, &c.—Concluded.

Date.	Names of Places of Observations.	Geographic Position.		Times of High Water.		Winds.		General Remarks.
		Lat. N.	Lon. W.	Observed.	Reduced to full & change.	Direction.	Force.	
1826.		° ′	° ′	h. m.	h. m.			
July 20	W. Horton River -	69 50	125 55	4 15 P.M.	3 15	W.N.W.	9	
21	- - - - -	- -	- -	5 0 A.M.	3 49	- -	7	
27	Cape Lyon - -	69 46	122 51	11 50 A.M.	6 33	E.N.E.	8	Stream of flood from the Eastward. Rise and fall 14 inches.
30	Three miles from Buchanan River	69 24	120 03	5 0 P.M.	8 20	W.N.W.	8	Ice close and heavy, Rise and fall 9 inches.
Aug, 1	Point Wise - -	69 03	119 00	8 30 P.M.	7 04	West.	4	Compact ice.
3	Stapylton Bay -	68 52	116 03	9 0 P.M.	8 22	East.	2	In a bay filled with ice.
4	Between C. Hope and C. Bexley	68 57	115 48	3 15 P.M.	8 25	E.S.E.	4	Ice to seaward.
5	Chantry Island - -	68 45	114 23	8 30 P.M.	7 22	W.S.W.	3	Loose masses of ice.
6	Seven miles from C. Krusenstern	68 32	113 53	9 00 P.M.	7 13	Variable.	- -	Flood from the S.E. Velocity 3 miles an hour.

CAPTAIN FRANKLIN'S NARRATIVE RESUMED.

CHAPTER VI.

Brief Notices of the Second Winter at Bear Lake—Traditions of the Dog-ribs—Leave Fort Franklin—Winter Journey to Fort Chipewyan—Remarks on the progress of improvement in the Fur Countries—Set out in Canoes on the Voyage Homeward—Join Dr. Richardson at Cumberland House—Mr. Drummond's Narrative—Arrival in Canada, at New York, and London.

Thursday, 21st.

DURING our absence on the sea-coast, Mr. Dease had employed the Canadians in making such repairs about the buildings as to fit them for another winter's residence, but he had not been able to complete his plans before the arrival of Dr. Richardson's party, through whose assistance they were finished shortly after our return. The inconvenience arising from the unfinished state of the houses was a trifle, when compared to the disappointment we felt at the poverty of our store, which contained neither meat nor dried fish, and the party was living solely on the daily produce of the nets, which, at this time, was barely sufficient for its support. Notwithstanding the repeated promises which the Fort hunters and the Dog-ribs in general had given us, of exerting themselves to collect provisions during the summer, we found that they had not supplied more than three deer since our departure. The only reason they assigned to Mr. Dease, on his remonstrating with them, was, that they had been withheld from hunting at any great distance from the Fort, by the fear of meeting the Copper Indians, who, they fancied, would be lying in wait to attack them. This excuse, however, had been so often alleged without a cause, that it was considered mere evasion, and we attributed their negligence to the indolence and apathy which mark the character of this tribe.

I need not dilate upon the anxieties which we felt at the prospect of commencing the winter with such a scanty supply of food. We at once sent off five men, provided with nets and lines, to the fishery in M'Vicar's Bay, which had been so productive in the preceding year, in the hope that, besides gaining their own subsistence, they might store up some fish for us, which could be brought to the Fort when the lake was frozen. Our anxiety was, in some measure, relieved on the 28th of September, by the arrival of Beaulieu and some hunters, from the north side of Bear Lake, with a supply of dried meat. The term of Beaulieu's engagement being now expired, he was desirous of quitting our service; and though he was our best hunter, Mr. Dease advised me to comply with his request, as he had collected a number of useless followers, whom we must have fed during the short days. He accordingly took his departure, accompanied by seventeen persons, which was a very important relief to our daily issue of provision. I furnished them with ammunition from the store to enable them to hunt on their way to Marten Lake, where they intended to fish until the return of spring.

October.

Calculating that the stores, which had been ordered from York Factory, must have arrived at Fort Norman, I despatched Mr. Kendall for them; and he returned on the 8th of October, with as much of them as his canoe would carry. The men were immediately furnished with warm clothing, of which the eastern party were in great need, having left every thing on quitting the sea-coast, except one suit each. We were rejoiced at the receipt of a large packet of letters from England, dated in the preceding February. They brought out the gratifying intelligence that my friend Lieutenant Back had been promoted, in December, 1825, to the rank of Commander. I likewise received a large packet of news papers from his Excellency the Earl of Dalhousie, Governor-in-Chief of Canada, to whom I take this opportunity of returning my best thanks for the warm interest he took in the welfare of the Expedition.

I shall now briefly trace the advance of winter: the nights were frosty and the weather was unsettled and gloomy, from the time of our arrival to the close of September. Heavy rain fell on the 2nd of October, which on the following day was succeeded by hard frost and much snow. The snow which fell on the 8th remained on the ground for the rest of the season. The small lake was frozen on the 12th, from which day we dated the commencement of winter as we had done in the preceding year. There was a succession of gales, and almost constant snow from that time to the close of the month; and on the 30th the thermometer first descended below zero. The snow then was much deeper than at the close of November in the former year. The last of the migratory birds, which were a few hardy ducks, took their departure on the 18th of October.

November. Stormy weather kept the Bear Lake open until the 16th of November, nine days later than the year before; and for some weeks we received no assistance from the nets, which again reduced our stock of meat to a small quantity. The same occupations, amusements, and exercise, were followed by the officers and men as in the former residence; and the occurrences were so similar, that particular mention of them is unnecessary. On the 25th of November we despatched some men with dogs and sledges to bring the remainder of the stores from Fort Norman. As it was my intention, as soon as the maps and drawings could be finished, to proceed on the ice to Fort Chipewyan, in order to secure provisions for the out-going of the party, and to reach England by the earliest conveyance, I requested of Mr. Brisbois to provide a cariole, sledges, and snow-shoes, for my journey, the birch of which they are made being plentiful in the neighbourhood of Fort Norman, and he having a better workman than any at our establishment. On the 28th Mackenzie arrived from M'Vicar's Bay, with an acceptable supply of fine white-fish. We learned from him that our party, as well as the Indians, were living in abundance; and that the latter had shown their wisdom this season, not only in taking up their quarters at that place, instead of remaining about the Fort, as they had done in the former year, but also in building themselves houses like those of our men, and thus having more comforts and better shelter than they had ever before enjoyed. The fishery opposite the Fort was now sufficiently productive for our wants, though the fish, from being out of season, disagreed so much with several of the men as to cause great debility, which was the more distressing to us, as we were unable to supply the indvalids with meat on more than two days in the week. Contrary to what had happened last season, we did not receive meat this year from more than six or seven persons of either the Hare Indians or Dog-rib tribes, after the ice set in; this happened, probably, from our being now unprovided with goods to exchange for their furs; though they had been expressly told in the spring, that we should have abundance of ammunition, tobacco, and other supplies, to purchase all the meat they would bring.

By the return of our men from Fort Norman, we learned that one of our Dog-rib hunters had murdered a man of his tribe, in the autumn, near the mouth of the Bear Lake River. The culprit being at the house, we inquired into the truth of the report, which was found correct; and he was in consequence instantly discharged from our service. His victim had been a man of notoriously loose habits, and in this instance had carried off the hunter's wife and child, while he was in pursuit of deer, at a great distance from the Fort. The husband pursued the guilty pair the moment he discovered their flight, and, on overtaking them, instantly shot the seducer; but the woman escaped a similar fate, by having the presence of mind to turn aside the muzzle of the gun when in the act of being discharged. She did not, however, escape punishment: her husband struck her senseless to the ground with the stock of his gun, and would have completed her destruction, but for the cries and intreaties of their only child. This transaction adds another to the melancholy list of about thirty murders which have been perpetrated on the borders of this lake since 1799, when the first trading post was established.

The Dog-rib Indians, being derived from the same stock with the Chipewyans, have many traditions and opinions in common with that people. I requested Mr. Dease to obtain answers from the old men of the tribe to a few queries which I drew up, and the following is the substance of the information he procured, which may be compared with the more extended statements by Hearne and Mackenzie, of the general belief of the Chipewyans.

The *first man*, they said, was, according to the tradition of their fathers, named Chapewee. He found the world well stocked with food, and he created children, to whom he gave two kinds of fruit, the black and the white, but forbade them to eat the black. Having thus issued his commands for the guidance of his family, he took leave of them for a time, and made a long excursion for the purpose of conducting the sun to the world. During this, his first absence, his children were obedient, and ate only the white fruit, but they consumed it all; the consequence was, that when he a second time absented himself to bring the moon, and they longed for fruit, they forgot the orders of their father, and ate of the black, which was the only kind remaining. He was much displeased on his return, and told them that in future the earth would produce bad fruits, and that they would be tormented by sickness and death—penalties which have attached to his descendants to the present day. Chapewee himself lived so long that his throat was worn out, and he could no longer enjoy life; but he was unable to die, until, at his own request, one of his people drove a beaver-tooth into his head.

The same, or another Chapewee (for there is some uncertainty on this head,) lived with his family on a strait between two seas. Having there constructed a weir to catch fish, such a quantity were taken, that the strait was choked up, and the water rose and overflowed the earth. Chapewee embarked with his family in a canoe, taking with them all manner of birds and beasts. The waters covered the earth for many days, but, at length, Chapewee said, we cannot live always thus, we must find land again, and he accordingly sent a beaver to search for it. The beaver was drowned, and his carcase was seen floating on the water; on which Chapewee despatched a musk-rat on the same errand. The second messenger was long absent, and when he did return was near dying with fatigue, but he had a little earth in his paws. The sight of the earth rejoiced Chapewee, but his first care was about the safety of his diligent servant, the rat, which he rubbed gently with his hands, and cherished in his bosom, until it revived. He next took up the earth, and mouldering it with his fingers, placed it on the water, where it increased by degrees until it formed an island in the ocean. A wolf was the first animal Chapewee placed on the infant earth, but the weight proving too great, it began to sink on one side, and was in danger of turning over. To prevent this accident the wolf was directed to move round the island, which he did for a whole year, and in that time the earth increased so much in size, that all on board the canoe were able to disembark on it. Chapewee, on landing, stuck up a piece of wood, which became a fir-tree, and grew with amazing rapidity, until its top reached the skies. A squirrel ran up this tree, and was pursued by Chapewee, who endeavoured to knock it down, but could not overtake it. He continued the chase, however, until he reached the stars, where he found a fine plain, and a beaten road. In this road he set a snare made of his sister's hair, and then returned to the earth. The sun appeared as usual in the heavens in the morning, but at noon it was caught by the snare which Chapewee had set for the squirrel, and the sky was instantly darkened. Chapewee's family on this said to him, you must have done something wrong when you were aloft, for we no longer enjoy the light of day; "I have," replied he, "but it was unintentionally." Chapewee then endeavoured to repair the fault he had committed, and sent a number of animals up the tree to release the sun, by cutting the snare, but the intense heat of that luminary reduced them all 'to ashes. The efforts of the more active animals being thus frustrated, a ground mole, though such a grovelling and awkward beast, succeeded by burrowing under the road in the sky, until it reached and cut asunder the snare which bound the sun. It lost its eyes, however, the instant it thrust its head into the light, and its nose and teeth have ever since been brown, as if burnt. Chapewee's island, during these transactions, increased to the present size of the American Continent; and he traced the course of the rivers, and scraped out the lakes by drawing his fingers through the earth. He next allotted to the quadrupeds, birds, and fishes, their different stations, and endowing them with certain capacities, he told them that they were in future to provide for their own safety, because man would destroy them whenever he found their tracks; but to console them, he said, that when they died they should be like a seed of grass, which, when thrown into the water, springs again into life. The animals objected to this arrangement, and said, let us when we die be as a stone which, when thrown into a lake, disappears for ever from the sight of man. Chapewee's family complained of the penalty of death entailed upon them for eating the black fruit, on which he granted that such of them as dreamed certain dreams should be men of medicine, capable of curing diseases and of prolonging life. In order to preserve this virtue, they were not to tell their dreams until a certain period had elapsed. To acquire the power of foretelling events, they were to take an ant alive, and insert it under the skin of the palm of the hand, without letting any one know what they had done.

For a long time Chapewee's descendants were united as one family, but at length some young men being accidentally killed in a game, a quarrel ensued, and a general dispersion of mankind took place. One Indian fixed his residence on the borders of the lake, taking with him a dog big with young. The pups in due time were littered, and the Indian, when he went out to fish, carefully tied them up to prevent their straying. Several times as he approached his tent, he heard a noise of children talking and playing; but on entering it he only perceived the pups tied up as usual. His curiosity being excited by the noises he had heard, he determined to watch, and one day pretending to go out and fish, according to custom, he concealed himself in a convenient place. In a short time he again heard voices, and rushing suddenly into the tent, beheld some beautiful children sporting and laughing, with the dog-skins lying by their side. He threw the skins into the fire, and the children, retaining their proper forms, grew up, and were the ancestors of the Dog-rib nation.

On Mr. Dease questioning some of the elderly men as to their knowledge of a supreme Being, they replied—" We be-" lieve that there is a Great Spirit, who created every thing, " both us and the world for our use. We suppose that he dwells " in the lands from whence the white people come, that he is " kind to the inhabitants of those lands, and that there are " people there who never die: the winds that blow from that " quarter (south) are always warm. He does not know the " wretched state of our island, nor the pitiful condition in " which we are."

To the question, whom do your medicine men address when they conjure? They answered,—" We do not think that they " speak to the master of life, for if they did, we should fare " better than we do, and should not die. He does not inhabit " our lands."

December. On the evening of the 1st of December a brilliant comet appeared in the westernquarter, which had been indistinctly seen the two preceding nights. A line drawn through α and η Ursæ Majoris led to its position; it also formed a trapezium with α Aquilæ and α Lyræ and α Corona Borealis. This was the last night of its being visible. The temperature had been unusually high for several days, about this time + 18 above zero; and, with the exception of the night of the 1st, the atmosphere gloomy; and we amused ourselves with conjecturing, whether this extrordinary warmth, and the density of the clouds, could in any way be ascribed to the comet.

At Christmas we were favoured by a visit from Mr. Brisbois, to whom we felt much obliged for the care he had taken of our sea-stores, beside many personal civilities. The visit of a stranger is always heartily welcomed in such a desolate region, and to provide for the entertainment of the party during Mr. Brisbois's stay, Captain Back and Mr. Kendall displayed their ingenuity in cutting out several pasteboard figures, to represent behind an illuminated screen the characters of a comic piece, which Captain Back had written for the occasion. The exhibition was entirely new to most of the party, and its execution afforded such general amusement, that it was repeated on three nights at the request of the men. The New Year was celebrated by a dance, which closed our festivities; and on Mr. Brisbois quitting us the following day, we resumed our ordinary occupations. Two Hare Indians arrived at the fort, whom Mr. Kendall recognised as the persons who had brought provisions to Dr. Richardson's party, as soon as they had heard of his having reached the Bear Lake

January.

Portage; and we had much pleasure in rewarding their promptitude on that occasion, by a substantial present and a silver medal. They were particularly pleased at the medals, and assured us that they should be proud to show them to the rest of their tribe as tokens of our approbation.

On the evening of the 4th of January, the temperature being —52, 2°, Mr. Kendall froze some mercury in the mould of a pistol bullet, and fired it against a door at the distance of six paces. A small portion of the mercury penetrated to the depth of one eighth of an inch, but the remainder only just lodged in the wood. Much snow fell in the second week of January; and on the 12th, we ascertained that its average depth was two feet in the sheltered parts of the woods. The weather became mild after the 20th; and on the 22nd, the sun's rays were so powerful as to raise a spirit thermometer with a blackened bulb, to + 30, 5°. when the temperature of the air was — 3, 5°. A very brilliant and clearly defined parhelion was visible at the time, and there were only a few light clouds. The wind was east, and as usual, with the wind from that quarter when the sky is clear, the distant land appeared much distorted by refraction.

The documents which had been preparing being now nearly finished, we sent for the cariole, &c. from Fort Norman. When the men came back, they brought the information, that, according to the report of the Indians, the ice was so rough on the Mackenzie above Fort Norman, that travelling would be extremely difficult. I therefore abandoned the intention of proceeding by that way, and resolved on passing through the woods to Fort Simpson, as soon as guides could be procured. The delay afforded me the opportunity of registering the lowest temperature we had witnessed in this country. At a quarter after eight in the morning of the 7th of February, the thermometer descended to —58°; it had been —57. 5°, and 57. 3° thrice in the course of this and the preceding day—between the 5th and 8th, its general state was from —48° to —52°, though it occasionally rose to —43°.

18th.

February.

At Fort Enterprise, during a similar degree of cold, the atmosphere had been calm: but here we had a light wind, which sometimes approached to a fresh breeze. The sky was cloudless the whole time. Some of our men, as well as the Indians, were travelling on the lake during this cold without experiencing any greater inconvenience than having their faces frost bitten. The dogs, however, suffered severely, three being completely lamed by the frost, and all of them becoming much thinner.* These cold days were followed by windy though mild weather, which brought the rein-deer nearer to the Establishment; and our hunters killed seven within a day's march. Their re-appearance in our neighbourhood was very gratifying to the whole party, as we were heartily tired of a fish-diet, and I felt an especial pleasure at being able to quit the place without the least apprehension of the party being in want of provision.

The following is a list of the amount of provision we obtained at Fort Franklin, from the time of Mr. Dease's arrival to the close of January 1827; independent of the supplies of pemmican, &c. for the sea voyage, which were procured from the Hudson's Bay Company.

Small Fish, Bear Lake Herring, 79,440.—Trout, 3,475.—Pounds of fresh meat, 24,053.—Dried ribs of Rein Deer, 2,370.—Pounds of pounded deer's meat, 1,744.—Pounds of fat or tallow, 2,929.—Rein-deer tongues, 1,849.—Beaver, 12.—Partridges, 386.—Hares, 52.

On the 16th of February, Augustus and two Dog-ribs were sent forward to be at the track in the line of my intended route. My departure being fixed for the 20th, the charts, drawings, journals, and provisions were distributed between the cariole and three sledges of which my train consisted; and as the dogs were in too weak a condition for drawing heavy burdens, two Indians were engaged, to accompany us four days, for the purpose of carrying part of the pemmican. I afterwards delivered written instructions to Captain Back, directing him to proceed to York Factory as soon as the ice should break, and from thence, by the Hudson's Bay ship, to England, taking with him the British party, but to send the Canadians to Montreal. Augustus and Ooligbuck were to be forwarded to Churchill, that they might rejoin their relatives.

Tuesday, 20th. At ten A.M., I quitted the Fort, accompanied by five of our men and the two Indians, the latter dragging each sixty pounds of pemmican on their sledges. Captain Back, the officers, and men assembled to give us a farewell salute of three hearty cheers, which served to renew my regret at leaving a society whose members had endeared themselves to me by unremitting attention to their duties, and the greatest personal kindness. We crossed the lake expeditiously, favoured by a north-west gale, and then continued our

* Notwithstanding the severity of the weather, we had great difficulty in causing these animals to depart from their usual custom of sleeping in the snow, and in inducing them to occupy the warm houses which were built for them.

course to the southward until sunset. The mode of bivouacking in the winter, as well as the course of proceeding, having been so fully described in my former Narrative, and by several other travellers in this country, I need not repeat them. We usually set forward at the first appearance of light and marched until sunset, halting an hour to breakfast. The rate of walking depended on the depth of snow; where the track was good, we made about two miles in the hour.

On the evening of the second day, we were deserted by our Indian companions, who, as we afterwards learned, took advantage of the rest of the party being some distance in advance of them, to turn back to the nearest wood, and there deposit the pemmican on a stage which they constructed by the road side. Supposing that they had only halted in consequence of the gale that was then blowing, we did not send to look after them before the following morning, when every trace of their path was covered with the snow drift; and as I considered we might possibly spend some time in a fruitless search, I thought the wisest course was to put the party and dogs on a shorter allowance than usual, and proceed on our journey. Their conduct affords another instance of the little dependence that ought to be placed on the Indians of this country, when more than ordinary exertion is required.

We travelled fifty miles through a swampy level country, thinly wooded, with a few ridges of hills visible in the distance, east and west of our course. The country was uneven and better wooded for the succeeding thirty miles. We next crossed a steep range of hills elevated about eight hundred feet above the surrounding land, and then passing over a succession of lower hills and vallies, descended to the Mackenzie, and following that river for thirty miles, came to Fort Simpson on the 8th of March; the whole distance being two hundred and twenty miles, and for the last one hundred and seventy miles, through a well wooded country. We crossed several rivers which flow into the Mackenzie, and some considerable lakes which are laid down in the map. But one solitary family of Indians were seen on the journey, and these were stationed within a day's march of Fort Simpson. They had inclosed large tracts of ground with hedges, in which they set snares for hares, and, being very successful, were living in abundance, and were well clothed, their dress consisting principally of hare skins.

March.

As soon as Mr. Smith, the chief Factor of the District, was informed of our approach, and that we were short of provisions, in consequence of the Indians having made off with the pemmican, he kindly sent a supply of fresh meat for our use; and on our arrival at the Fort, he gave us the most friendly reception. Our Indian guide had never been nearer to Fort Simpson by land, than the Lake of the Elevated Land, and only once by the course of the Mackenzie, many years before the Fort was built; and yet if he had not been led aside by falling upon the track leading to the Indians above-mentioned, he would have come upon the Mackenzie, directly opposite Fort Simpson. His course he told me was governed by his recollection of a particular mountain, which he remembered to have noticed from the Mackenzie, and which we now passed within two miles, but on his former visit, he did not approach it nearer than eighteen miles. Its outline must have appeared so different when seen from these distances, that one can hardly imagine a less observant eye than that of an Indian recognising any of its distinguishing points, especially as it was not a detached mountain, but formed one of a line of hills of considerable extent. Our dogs being completely tired, I remained a week to recruit their strength. During this interval I had the opportunity of examining all the accounts which the Hudson's Bay Company had to present for supplies to the Expedition from this department, and of making provision for the outward journey of Captain Back and his party. Arrangements were also made, that the Hudson's Bay Company should take, at a valuation, the spare stores of the Expedition on its quitting Bear Lake. I accompanied Mr. Smith to a part of the River of the Mountains, where a portion of the bank, several acres in extent, had been torn off, and thrown a considerable distance into the channel of the river. The disruption took place in the preceding November, some days after the water had been frozen, and when there was no apparent cause for its separation. When the water is flowing over the banks, and the earth is in consequence loosened, the falling of the bank is not unfrequent in the Mackenzie, though on a much smaller scale than in this instance. I can only account for the separation of the mass after the ground had been frozen, by the supposition, that there was some spring of warm water in its rear, which loosened the soil, and that the pressure of the ice contributed, with the weight of snow at the top, to its overthrow.

At the time of my visit, an Indian woman committed suicide, by hanging herself, in a fit of jealousy, at an encampment, a short distance from the Fort. I had thought that suicide was extremely rare among the Northern Indians; but I subsequently learned that it was not so uncommon as I had imagined, and I was informed of two instances that occurred in the year of 1826. The weather was remarkably mild; during my stay icicles were formed on the southern front of the house, and there were many other indications of an early spring.

On the afternoon of the 15th of March I took leave of Mr. Smith, who kindly furnished me with his best dog for my cariole, one of mine having proved unfit for the journey to Slave Lake; we were also indebted to him for the skin of a mountain goat and a lynx; and to Mr. M'Pherson for the skins of several smaller animals and birds, from the neighbourhood of the Rocky Mountains, which they added to our collection. Having sent back one of my men with the Indian guides to Bear Lake, we had now only two sledges; but as we were unable to carry the whole of our lading, Mr. Smith had the goodness to send a sledge and one of his men to convey a part of the provisions for four days. At the distance of eight miles we met two men with a cariole and sledge, which Mr. M'Vicar had sent for my use from Slave Lake; but being well provided I did not require the services of this party, though we derived great benefit from their track as we proceeded, and also from some deposits of provision which they had made on the route.

Thursday, 15th.

Following the course of the Mackenzie, we arrived, on the 21st, at the expansion of the river called the Little Lake, and there had the pleasure of meeting two Canadians, on their way to Bear Lake, with a packet of letters from England. We hastened towards the shore and encamped; and though the night was piercingly cold, I spent the greatest part of it most agreeably, scanning the contents of the box by the unsteady light of a blazing fire. After breakfast next morning I despatched the packet to its destination, under the charge of M'Leay, who had accompanied me from Bear Lake, and retained one of the Canadians in his stead. We arrived at Fort Resolution, on the Slave Lake to breakfast, on the 26th, and I once more had the happiness of receiving the friendly attentions of Mr. M'Vicar, to whom it will be remembered by the readers of my last Narrative, that the members of that Expedition were so greatly indebted for his tender care of them after their sufferings. Dr. Richardson had quitted this place in the preceding December, for the purpose of joining Mr. Drummond, the Assistant Botanist in the Saskatchawan River, and that he might have the benefit of an earlier spring than in this quarter to collect plants. The prospect here being completely wintry, I made another halt of eight days, being desirous of remaining as long as I could,

Wednesday, 21st.

without incurring the risk of exposure to the thaw on my way to Fort Chipewyan.

I was glad to find that the Chipewyans and Copper Indians were at length employing dogs to drag their sledges. A superstitious belief that their own origin was derived from those animals, had for several years past thrown this laborious and degrading occupation on the poor women, who, by the change, experienced a most happy relief. It was indeed, highly gratifying to observe that these Indians no longer beat their wives in the cruel manner to which they had been formerly accustomed; and that, in the comparative tenderness with which they now treat the sex, they have made the first and greatest step to all moral and general improvement.

It will be recollected that on receiving, at Bear Lake, a report of the traces of white people having been seen near the sea-coast, I had requested that Mr. M'Vicar would collect a party of Indians, and send them to the spot to convey a letter from me to Captain Parry. Mr. M'Vicar now informed me that some Indians had left his Fort for the purpose, under the charge of a Canadian, named Joseph St. Pierre, who volunteered for the occasion, but the Indians continued with him only for a short distance beyond the east end of Slave Lake, when they became weary of their journey, and dropping off one by one, left him alone. St. Pierre, however, having determined to deliver the letter to Captain Parry, if possible, persevered for many days in a fruitless search for the river on the banks of which the marks were reported to have been seen; even after he had sustained the loss of all his clothes (except those on his person,) by the grass catching fire when he was asleep; but at length, being short of food, his shoes worn out, and almost without covering for his feet, he was compelled to return to the Fort. He was not at the house at the time of my visit, but I left an order with Mr. M'Vicar, that he might be rewarded for his zeal and exertions, and handsomely remunerated for his loss.

April. The subsequent journey to the Athabasca Lake occupied eight days; we arrived at Fort Chipewyan in the afternoon of the 12th of April. I found Mr. Stewart, the Chief Factor of the Department, surrounded by a large body of Indians, who quitted the Fort as soon as they had exchanged their furs, in order to seek their living by fishing and hunting wild fowl, instead of passing four or five weeks in indolence about the Establishment, as had been their custom at this season for many preceding years. This beneficial change of conduct, on their part, is owing to the Hudson's Bay Company having ceased to bring spirits into the northern department; and to some other judicious regulations which the Directors have made respecting the trade with the natives. The plans now adopted offer supplies of clothes, and of every necessary, to those Indians who choose to be active in the collection of furs; and it was pleasing to learn, that the natives in this quarter had shown their acquiescence in these measures by increased exertion during the preceding winter. Some other very wholesome regulations have been introduced by the Company; amongst others, the Sabbath is ordered to be properly observed, and Divine Service to be read at every post. They have also directed, where the soil will allow, a portion of ground to be cultivated for the growth of culinary vegetables at each of their establishments, and I witnessed the good effects of this order, even at this advanced post, where the ground is rocky; the tables of the officers being supplied daily, and those of the men frequently, with potatoes and barley. Such luxuries were very rarely found beyond Cumberland House, on the route that we travelled during my former journey.

Feeling a deep interest in the welfare of this country, in which I have spent a large portion of the last seven years, I have much pleasure in recording these improvements; und in stating my conviction, that the benevolent wishes of the Directors, respecting the inhabitants of their territories, will be followed up with corresponding energy by the resident Governor, the chief factors, and the traders of the Company.

I mentioned in my former Narrative, that the Northern Indians had cherished a belief for some years, that a great change was about to take place in the natural order of things, and that among other advantages arising from it, their own condition of life was to be materially bettered. This story, I was now informed by Mr. Stewart, originated with a woman, whose history appears to me deserving of a short notice. While living at the N.W. Company's Post, on the Columbia River, as the wife of one of the Canadian servants, she formed a sudden resolution of becoming a warrior; and throwing aside her female dress, she clothed herself in a suitable manner. Having procured a gun, a bow and arrows, and a horse, she sallied forth to join a party of her countrymen then going to war; and, in her first essay, displayed so much courage as to attract general regard, which was so much heightened by her subsequent feats of bravery, that many young men put themselves under her command. Their example was soon generally followed, and, at length she became the principal leader of the tribe, under the designation of the "Manlike Woman." Being young, and of a delicate frame, her followers attributed her exploits to the possession of supernatural power, and, therefore, received whatever she said with implicit faith. To maintain her influence during peace, the lady thought proper to invent the above-mentioned prediction, which was quickly spread through the whole northern district. At a later period of her life, our heroine undertook to convey a packet of importance from the Company's Post on the Columbia to that in New Caledonia, through a tract of country which had not, at that time, been passed by the traders, and which was known to be infested by several hostile tribes. She chose for her companion another woman, whom she passed off as her wife. They were attacked by a party of Indians, and though the Manlike Woman received a wound in the breast, she accomplished her object, and returned to the Columbia with answers to the letters. When last seen by the traders, she had collected volunteers for another war excursion, in which she received a mortal wound. The faith of the Indians was shaken by her death, and soon afterwards the whole of the story she had invented fell into discredit.

In the Athabasca department, which includes Slave Lake and Peace River, as well as in the more southern districts, the autumn of 1826, and the following winter, were unusually mild. Near the Saskatchawan River, there was so little snow before the middle of January, that the sledges could not be used; but at Bear Lake, and throughout the Mackenzie, the weather was severe during the same periods, and the snow came early; hence it would appear, that even in this climate the meteorological register kept at any one place, affords no index from whence we can judge of the season at another. In my journey from Slave Lake to the Athabasca we had a snow-storm for three days, which we found did not extend beyond sixty miles; and on our arrival at Fort Chipewyan, we learned there had not been a single shower during these days. The only coinciding circumstance, at the different stations this year, was the prevalence of north-east winds.

Sunday, 15th. We welcomed the appearance of two of the large-sized swans on the 15th April, as the harbingers of spring; the geese followed on the 20th; the robins came on the 7th May; the house martins appeared on the 12th, and in the course of a week were busily employed repairing their nests; and the barn or forked-tail swallows arrived on the 20th; and on the same day, the small-sized swans were seen, which the traders consider the latest of the migratory birds.

The only symptoms of reviving vegetation at this period, were a few anemones in flower, and the bursting of some catkins of willows; but we learned by an arrival of a boat from the Peace River that, even so early as the 14th, the trees were in full foliage at not more than a day's journey from the lake. The barley was sown at Fort Chipewyan on the 15th May, potatoes on the 21st, and the garden seeds on the 22d, which were expected to be ready for use by the close of the following September. As an experiment, whether the barley would yield a better crop by remaining in the ground through the winter, some had been sown in the preceding autumn, but only a few of the plants appeared at the close of this month, and the crop did not promise favourably.

May, 20th.

Some canoes having arrived on the 26th of May with the furs from Slave Lake, the last of the Company's brigade of boats was despatched to York Factory. Augustus, who was desirous of seeing Dr. Richardson again before his departure from the country, and two other men of the Expedition, embarked in them. I embarked on the 31st May in the Company's light canoe with Mr. Stewart and Mr. M'Vicar, having previously made the necessary arrangements for the passage of Captain Back and his party. We reached Cumberland House on the 18th June, where I had the happiness of meeting Dr. Richardson after a separation of eleven months. I learned from him that during our absence in the north, Mr. Drummond the Assistant Botanist had been indefatigable in collecting specimens of Natural History, having been sent for that purpose to the Rocky Mountains at the head of the Athabasca River; in the course of which service, he had been exposed to very great privations. To his perseverance and industry, science is indebted for the knowledge of several new and many rare quadrupeds, birds, and plants. That the reader may form some notion of the labour he sustained, and the zeal he displayed in making his very valuable and highly interesting collections, and to point out to the naturalist, the districts from whence they were brought, I subjoin his brief account of his journey in his own words.

1825, June.

"I remained at Cumberland House about six "weeks after the departure of Captain Back and "Mr. Kendall, in June, 1825, when the Company's boats with "the brigade of traders for the Columbia, arriving from York "Factory, I accompanied them up the Saskatchawan River "two hundred and sixty miles to Carlton House. The unset-"tled state of the Indians in that neighbourhood rendering ex-

"cursions over the plains very unsafe, I determined on pro-
"ceeding with the brigade as far as the Rocky Mountains.
"We left Carlton House on the 1st of September, and reached
"Edmonton, which is about four hundred miles distant on the
"20th of the same month. Sandy plains extend without ma-
"terial alteration the whole way, and there is, consequently,
"little variety in the vegetation; indeed, I did not find a single
"plant that I had not seen within ten miles of Carlton House,
"although I had an opportunity of examining the country
"carefully, having performed the greater part of the journey
"on foot. After a halt of two days at Edmonton, we conti-
"nued our route one hundred miles farther to Fort Assinaboyn
"on the Red Deer River, one of the branches of the
"Athapescow. This part of the journey was performed
"with horses through a swampy and thickly wooded country,
"and the path was so bad, that it was necessary to reduce the
"luggage as much as possible. I therefore took with me only
"one bale of paper for drying plants, a few shirts, and a blan-
"ket; Mr. M'Millan, one of the Company's chief traders, who
"had charge of the brigade, kindly undertaking to forward the
"rest of my baggage in the ensuing spring. We left Fort
"Assinaboyn to proceed up the Red Deer River to the Moun-
"tains, on the 2d of October; but the Canoe ap-
"pointed for this service being very much lum-
"bered, it was necessary that some of the party should travel
"by land, and of that number, I volunteered to be one. A
"heavy fall of snow, on the third day after setting out, ren-
"dered the march very fatiguing, and the country being thickly
"wooded and very swampy, our horses were rendered useless
"before we had travelled half the distance.

October, 2d.

"We reached the mountains on the 14th, and I continued
"to accompany the brigade, for fifty miles of the Portage-road,
"to the Columbia, when we met a hunter whom Mr. M'Millan
"hired to supply me with food during the winter. The same
"gentleman having furnished me with horses and a man to
"take care of them, I set out with the hunter and his family
"towards the Smoking River, one of the eastern branches of
"the Peace River, on which we intended to winter. My guide,
"however, loitered so much on the way, that the snow became
"too deep to admit of our proceeding to our destination, and
"we were under the necessity of leaving the Mountains alto-
"gether, and taking up our winter-quarters about the end of
"December, on the Baptiste, a stream which falls
"into the Red Deer River. During the journey, I
"collected a few specimens of the birds that pass the winter

December.

" in the country, and which belong principally to the genera
" *tetrao* and *strix*. I also obtained a few mosses, and on Christ-
" mas day, I had the pleasure of finding a very minute *gym-*
" *nostomum*, hitherto undescribed.

" In the winter, I felt the inconvenience of the want of my
" tent, the only shelter I had from the inclemency of the wea-
" ther being a hut built of the branches of trees. Soon after
" reaching our wintering ground, provisions became very scarce,
" and the hunter and his family went off in quest of animals,
" taking with them the man who had charge of my horses to
" bring me a supply as soon as they could procure it. I re-
" mained alone for the rest of the winter, except when my
" man occasionally visited me with meat; and I found the
" time hang very heavy, as I had no books, and nothing could
" be done in the way of collecting specimens of Natural His-
" tory. I took however, a walk every day in the woods to
" give me some practice in the use of snow shoes. The winter
" was very severe, and much snow fell until the end of March,
" when it averaged six feet in depth; in consequence of this,
" I lost one of my horses, and the two remaining ones became
" exceedingly poor. The hunter was still more unfortunate,
" ten of his young colts having died.

April 1826.

" In the beginning of April, 1826, setting out for the
" Columbia Portage road, I reached it after a fa-
" tiguing march on the sixth day, and two days afterwards,
" had the pleasure of meeting Mr. M'Millan, who brought me
" letters from Dr. Richardson, informing me of the welfare of
" the Expedition; and he also placed me in comparatively
" comfortable circumstances by bringing my tent, a little tea
" and sugar, and some more paper. I remained on the Por-
" tage preparing specimens of birds until the 6th of
" May, when the brigade from the Columbia arrived.
" On that day the *Anemone cuneifolia*, and *Ludoviciana* and
" *Saxifraga oppositifolia*, began to flower in favourable situa-
" tions. My hunter, who had, in the mean time, returned to
" our late wintering ground, now sent me word that he had
" changed his mind, and would not accompany me into the
" Mountains, as he had engaged to do. His fickleness deranged
" my plans, and I had no alternative but to remain with the
" man who had charge of the horses used on the Columbia
" Portage, and botanize in that neighbourhood.

May, 6th.

August.

" On the 10th of August, I set out with another
" hunter, upon whom I had prevailed to conduct me
" to the Smoking River, although, being disappointed in a sup-
" ply of ammunition, we were badly provided. We travelled

"for several days without meeting with any animals, and I "shared the little dried provision which I had with the hunter's "family. On the 15th we killed a Mountain sheep, which "was quickly devoured, there not being the smallest appre- "hension at the time that famine would overtake us—day after "day, however, passed away without a single head of game of "any description being seen, and the children began to com- "plain loudly; but the hunter's wife, a young half-breed woman, "bore the abstinence with indifference, although she had two "infant twins at the breast. On the 21st, we found two young "porcupines, which were shared amongst the party, and two "or three days afterwards, a few fine trout were caught. We "arrived in the Smoking River on the 5th of September, "where the hunter killed two sheep, and a period was put to "our abstinence, for before the sheep were eaten, he shot "several buffaloes.

"We proceeded along the Mountains until the "24th of September, and had reached the head "waters of the Peace River, when a heavy fall of snow stop- "ped my collecting plants for that season. I was, however, "very desirous of crossing the Mountains to obtain some "knowledge of the vegetation on the Columbia River, and, "accordingly, I commenced drying provisions to enable me "to accompany the Columbia brigade, when it arrived from "Hudson's Bay. I reached the Portage on the 9th of Octo- "ber, and on the 10th the brigade arrived, and I re- "ceived letters from Captain Franklin, instructing "me to descend in the spring of 1827, time enough to rejoin "the Expedition on its way to York Factory. It was, there- "fore, necessary that I should speedily commence my return, "and having gone with the brigade merely to the west-end of "the Portage, I came back again on the 1st of November. "The snow covered the ground too deeply to permit me to "add much to my collections in this hasty trip over the Moun- "tains, but it was impossible to avoid remarking the great "superiority of climate on the western side of that lofty range. "From the instant the descent towards the Pacific commences, "there is a visible improvement in the growth of timber, and "the variety of forest trees greatly increases. The few mosses "that I gleaned in the excursion were so fine, that I could not "but deeply regret that I was unable to pass a season or two "in that interesting region.

September.

October.

"Having packed up all my specimens, I embarked on the "Red Deer River, with Mr. M'Donald, one of the Company's "officers, who was returning from a long residence on the

"Columbia with his family, and continued to descend the "stream until we were set fast by the frost. I then left Mr. "M'Donald in the charge of the baggage, and, proceeding on "foot to Fort Assinaboyn, for the purpose of procuring horses, "I reached it on the fifth day. It was several days before the "horses could be obtained, and they were several more in travelling from the Fort to Mr. M'Donald, during which time "that gentleman and his family were very short of provisions. "The relief, however, arrived opportunely, and they reached "the Fort in safety. After resting a few days, I set out for "Edmonton, where I remained for some months.

"The winter express brought me a letter from Dr. "Richardson, requesting me to join him at Carlton "House in April, and I accordingly set out for that place on "snow shoes, on the 17th of March, taking with me single "specimens of all the plants gathered on the Mountains, lest "any accident should happen to the duplicates which were to "come by canoe in the spring. Two men with a sledge drawn "by dogs accompanied me, but the Indian inhabitants of the "plains being very hostile, we made a large circuit to avoid "them, and did not reach Carlton House before the 5th of "April. We suffered much from snow-blindness on the "march, the dogs failed from want of food, we had to "carry the baggage on our backs, and had nothing to eat for "seven days. These sufferings were, however, soon forgotten "in the kind welcome I received from Dr. Richardson, and "Mr. Prudens, the Company's Chief Trader at Carlton, and "the hospitable entertainment and good fare of the latter gentleman's table enabled me speedily to recruit my lost strength.

March.

April.

"My collections on the Mountains amounted to about fifteen hundred species of plants, one hundred and fifty birds, "fifty quadrupeds, and a considerable number of insects."

There being yet two months in which Mr. Drummond might continue his researches, before Captain Back could arrive at Cumberland House, Dr. Richardson had left him on the Saskatchawan River.

June.

After remaining part of a day at Cumberland House, we proceeded on our journey, Dr. Richardson following in one of the Company's boats. I reached Norway house on the 24th of June, and Dr. Richardson on the third day after. Mr. Simpson, the resident Governor of the Company, was absent on urgent business at York Factory; but, previous to his departure, he had provided a canoe, and some additional men, with every other requisite for my journey. We found here Mr. Douglass, who had been sent to the Columbia River

18th.

by the Horticultural Society, as a Collector of Natural History, and who had recently crossed the Rocky Mountains, for the purpose of proceeding to England from Hudson's Bay. This gentleman being desirous of occupying himself previous to the arrival of the ship, in making an addition to his collection from the neighbourhood of the Red River Colony, I felt happy in being able to give him a conveyance, in the canoe with Dr. Richardson and myself, through Lake Winipeg, to Fort Alexander, where he met another canoe that was going to the colony.

On quitting Norway House we took leave of our worthy companion, Augustus, who was to wait there until Captain Back should arrive. The tears which he shed at our parting, so unusual in those uncultivated tribes, showed the strength of his feelings, and I have no doubt, they proceeded from a sincere affection; an affection which, I can venture to say, was mutually felt by every individual. With great regret he learned that there was no immediate prospect of our again meeting, and he expressed a very strong desire to be informed, if another Expedition should be sent to any of the northern parts of America, whether by sea or land; and repeatedly assured me, that he and Ooligbuck would be ready at any time to quit their families and their country, to accompany any of their present officers, wherever the Expedition might be ordered.*

We reached Fort Alexander on the 8th of July, and Mr. Douglass having left us, I was enabled to offer a passage, as far as Montreal, to Monsieur Picard, one of the clergymen attached to the Roman Catholic Mission at the Red River Colony. We arrived at Lachine, near Montreal, on the 18th of August, and were hospitably entertained by Mr. James Keith, Chief Factor, and Agent of the Hudson's Bay Company, with whom we remained five days, to settle the accounts of the Expedition. After I had paid my respects to his Excellency, the Earl of Dalhousie, Governor in Chief of Canada, we proceeded to New York by the way of Lake Champlain. In our passage through the United States, we received the same kind attentions we had before experienced; our personal baggage, and the collections of Natural History, were forwarded by the officers of the customs without examination, and every assistance we required was promptly rendered.

August.

September.

Having embarked, in the packet ship, on the 1st of September, we reached Liverpool on the 26th,

* I have pleasure in mentioning that, by permission of Government, the pay which was due to Augustus and Ooligbuck, has been delivered to the Directors of the Hudson's Bay Company, who have undertaken to distribute it to them annually, in the way suited to their wants.

after an absence of two years, seven months and a half. Captain Back, Lieutenant Kendall, and Mr. Drummond, with the rest of the British party, arrived at Portsmouth on the 10th of October. I then received the distressing intelligence of the death of two excellent men, on their homeward passage from Bear Lake to York Factory; Archibald Stewart, who died from consumption; and Gustavus Aird, who was drowned in consequence of his jumping out of the boat, in his exertions to save her, when she was hurrying down the Pelican Fall, in Slave River. Until this account reached me, I had cherished the hope that our Expedition would have terminated without my having to record a single casualty. The loss of these men was the more deeply felt by me, from their uniform, steady, obedient, and meritorious conduct, which I had repeated opportunities of observing and admiring, while they were my companions in the Lion, during the voyage along the coast.

I must be allowed to add, that in this long homeward journey, in which there were no fresh discoveries to be made, nor any of those excitements that relieve the monotony of constant labour, and in which they had to contend with a succession of dangerous rapids, there was the same masterly skill and exemplary conduct evinced by Captain Back and Lieutenant Kendall; and the same patient and ready obedience by the men*, which had marked their whole conduct, while more immediately under my own observation.

On my arrival in London, on the 29th of September, accompanied by Dr. Richardson, I had the honour of laying the charts and drawings before his Royal Highness the Lord High Admiral, and Mr. Secretary Huskisson; and, from the latter, I received directions to publish an account of our proceedings.

In concluding this Narrative, I feel it incumbent on me to offer a few remarks on the subject of a *North-West Passage*, which, though it has not been the immediate object of the enterprises in which I have been engaged, is yet so intimately connected with them, as to have naturally excited in my mind, a strong and permanent interest. It is scarcely necessary to remark, that the opinion I ventured to express in my former work, as to the practicability of the passage†, has been considerably strengthened by the information obtained during the

* I am happy to add, that those men who had been in His Majesty's service before the present Expedition, have been rewarded by promotion.

† See page 388.

present Expedition. The Northern Coast of America has now been actually surveyed from the meridian of 109° to 149½° west; and again by the exertions of Captain Beechey, in His Majesty's ship the Blossom, from Icy Cape eastward to about 156° west, leaving not more than fifty leagues of unsurveyed coast, between Point Turnagain and Icy Cape. Further, the delineation of the west side of Melville Peninsula, in the chart of Captain Parry's Second Voyage, conjoined with information which we obtained from the Northern Indians, fairly warrants the conclusion, that the coast preserves an easterly direction from Point Turnagain towards Repulse Bay; and that, in all probability, there are no insurmountable obstacles between this part of the Polar Sea and the extensive openings into the Atlantic, through Prince Regent Inlet and the Strait of the Fury and Hecla.

Whenever it may be considered desirable to complete the delineation of the coast of the American Continent, I conceive that another attempt should be made to connect Point Turnagain with the important discoveries of Captain Parry, by renewing the Expedition which was undertaken by Captain Lyon, and which, but for the boisterous weather that disabled the Griper, must have long since repaid his well known zeal and enterprize with discoveries of very great interest.

In considering the best means of effecting the North-West Passage in a ship, it has hitherto been impossible not to assent to the opinion so judiciously formed, and so convincingly stated, by Captain Parry, that the attempt should be made from the Atlantic rather than by Behring's Straits, because the enterprise is then commenced after a voyage of short duration, subject to comparatively few vicissitudes of climate, and with the equipments thoroughly effective. But important as these advantages are, they may, perhaps, be more than balanced by some circumstances which have been brought to light by our Expedition. The prevalence of north-west winds during the season that the ice is in the most favourable state for navigation, would greatly facilitate the voyage of a ship to the eastward, whilst it would be equally adverse to her progress in the opposite direction. It is also well known, that the coast westward of the Mackenzie is almost unapproachable by ships, and it would, therefore, be very desirable to get over that part of the voyage in the first season. Though we did not observe any such easterly current as was found by Captain Parry in the Fury and Hecla Strait, as well as by Captain Kotzebue, on his voyage through Behring's Straits; yet this may have arisen from our having been confined to the navigation of the flats close to the shore; but if such a current does exist throughout the Polar Sea, it is evident that it would materially assist a ship commencing the undertaking from the Pacific, and keeping in the deep water, which would, no doubt, be found at a moderate distance from the shore.

The closeness and quantity of the ice in the Polar Seas vary much in different years; but, should it be in the same state that we found it, I would not recommend a ship's leaving Icy Cape earlier than the middle of August, for after that period the ice was not only broken up within the sphere of our vision, but a heavy swell rolling from the northward, indicated a sea unsheltered by islands, and not much encumbered by ice. By quitting Icy Cape at the time specified, I should confidently hope to reach a secure wintering place to the eastward of Cape Bathurst, in the direct route to the Dolphin and Union Straits, through which I should proceed.* If either, or both, of the plans which I have suggested be adopted, it would add to the confidence and safety of those who undertake them, if one or two depôts of provisions were established in places of ready access, through the medium of the Hudson's Bay Company.

Arctic discovery has been fostered principally by Great Britain; and it is a subject of just pride that it has been prosecuted by her from motives as disinterested as they are enlightened; not from any prospect of immediate benefit to herself, but from a steady view to the acquirement of useful knowledge, and the extension of the bounds of science. Each succeeding attempt has added a step towards the completion of northern geography; and the contributions to natural history and science have excited a general interest throughout the civilized world. It is, moreover, pleasing to reflect that the loss of life which has occurred in the prosecution of these discoveries does not exceed the average number of deaths in the same population at home under circumstances the most favourable. And it is sincerely to be hoped that Great Britain will not relax her efforts until the question of a north-west passage has been satisfactorily set at rest, or at least until those portions of the northern shores of America, which are yet unknown, be laid down in our maps; and which with the exception of a small space on the Asiatic continent eastward of Shelatskoi Noss, are the only intervals wanting to complete the outline of Europe, Asia, and America.

* See Dr. Richardson's opinion in favour of this route, p. 218.

END OF THE NARRATIVE.

Summary of the Distances travelled by the Expedition, from its Landing in America until its Embarkation.

	Statute Miles.
Distance travelled in 1825, as given in page 60 - - - -	5,803
Dr. Richardson and Mr. Kendall's excursion on the ice to the eastern parts of Bear Lake, in the Spring of 1826 - - - -	359
Distance travelled by the Western Party in 1826 (given in p. 235.)	2,048
Distance travelled by the Eastern Party in 1826, after its separation from the Western Party - - - - - - - - -	1,455
Return from Fort Franklin to New York - - - - -	4,000
Captain Back and Lieutenant Kendall's journey to York Factory, after quitting Captain Franklin's route - - - - -	520
Distance travelled by the Expedition in going and returning, including the excursions of detached parties - - -	14,185
Number of miles surveyed and laid down in the maps, but not all included under the head of discoveries, because the routes have been traversed by Traders	5,000

APPENDIX.

TOPOGRAPHICAL AND GEOLOGICAL NOTICES,

BY

JOHN RICHARDSON, M.D., F.R.S., &c.

SURGEON AND NATURALIST TO THE EXPEDITION.

[*Read before the Geological Society.*]

A VERY limited portion of my time could be allotted to geological inquiries. For eight months in the year the ground in the northern parts of America is covered with snow; and during the short summer, the prosecution of the main object of the expedition rendered the slightest delay in our journey unadvisable. The few hours that could be stolen from the necessary halts, for rest and refreshment, were principally occupied in the collection of objects for the illustration of botany and zoology. It is evident, that an account of the rock formations, drawn up under such circumstances, cannot be otherwise than very imperfect; but I have been led to publish it from the belief that, in the absence of more precise information, even the slightest notice of the rocks of the extreme northern parts of the American continent would be useful to those employed in developing the structure of the crust of the earth; the more especially, as it is not probable that the same tract of country will soon be trod by an expert geologist. The specimens of rocks I obtained have been deposited in the Museum of the Geological Society, and are referred to in the ensuing pages by the numbers affixed to them. The notices are arranged nearly in the order of the route of the expedition, commencing with Great Bear Lake, where our winter quarters were situated.

GREAT BEAR LAKE.

GREAT Bear Lake is an extensive sheet of water, of a very irregular shape, being formed by the union of five arms or bays in a common centre. The greatest diameter of the lake, measuring about one hundred and fifty geographical miles, runs from the bottom of Dease Bay, which receives the principal feeding stream, to the bottom of Keith Bay, from whence the Bear Lake River issues, and has a direction from N.E. to S.W. The transverse diameter has a direction from N.W. by W. to S.E. by E., through Smith and M'Tavish Bays, and is upwards of one hundred and twenty miles in length. M'Vicar Bay, the fifth arm of the lake, is narrower than the others, and being a little curved at its mouth, appears less connected with the main body of water. The light bluish-coloured water of Great Bear Lake is every where transparent, and is particularly clear near some primitive mountains, which exist in M'Tavish Bay. A piece of white rag, let down there, did not disappear until it descended fifteen fathoms. The depth of water, in the centre of the lake was not ascertained; but it is known to be very considerable. Near the shore, in M'Tavish Bay, forty-five fathoms of line did not reach the bottom. Owing to the barometers supplied to the expedition having been broken in an early period of its progress, the height of the surface of Bear Lake above the Arctic Sea could not be ascertained; but it is, probably, short of two hundred feet.* If this supposition comes near the truth, the bottom of M'Tavish Bay is below the level of the sea, and towards the centre of the basin of the lake the depression is probably still greater. The great lakes, Huron, Michigan, and Superior, which discharge their waters into the St. Lawrence, are reported to sink three hundred feet below the level of the ocean; and the Lake of the Mountains, or Chipewyan Lake and Great Slave Lake,† through which the Mackenzie flows, have, it is highly probable, some portions of their beds below the sea level.

In the autumn of 1825, I coasted the western and northern shores of the Great Bear Lake; and in the spring of 1826, travelled on the ice along its eastern and southern arms, leaving no part of its shores unexamined on these two surveys, except the north side of M'Tavish Bay. I did not, however, on these occasions, make excursions inland.

PRIMITIVE ROCKS.—GREAT BEAR LAKE.

At the south-east corner of M'Tavish Bay, primitive rocks form a hilly range which, at the distance of a mile or two from

* This was estimated by allowing one foot descent per mile for Bear Lake River, whose length is seventy miles; and three inches per mile for the descent of Mackenzie River, from the junction of the former river to the sea, being a distance of five hundred miles.

† In our former journey, we sounded near the Rein-Deer Islands in Slave Lake, with sixty-five fathoms line, without reaching the bottom.

the shore, attains an elevation of eight hundred or one thousand feet. The steep face of the range forms the shore of the lake for fifteen miles, and perhaps further, on a direction from N.W. by W. to S.E. by E., and is prolonged on the latter bearing, at the back of the lower country lying towards Point Leith. The general form of the hills is obtuse-conical, in some instances approaching to dome-shaped. None of them rise much above the others, and the vallies between them are seldom wide or deep. At a distance, some of the masses of rock appear round-backed; and in certain points of view, the crest of the ridge seems to consist of mammillary peaks. On a nearer approach, the individual hills are found to be composed of rounded eminences, having summits, generally, of an oblong form, and consisting of smooth, naked rock. Small mural precipices are frequent, and many detached blocks of stone lie beneath them. Between the eminences, there are level spots destitute of vegetation, and covered with small stones or gravel not much worn. A considerable portion of the gravel is granite or quartz, the debris, perhaps, of the rocks, of which the hills consist; it contains also some pieces of slate, and not a few of quartzose sandstone, neither of which I observed *in situ*. In the course of a walk of two miles over these hills, the only rock I observed was granite, verging in a few places towards gneiss, and generally whitish, with black mica. Sometimes the felspar is brownish-red, and the rock not unfrequently contains disseminated augite? The weathered surface of the stone was every where of a brick-red colour. In many spots the rocks split into such thin slaty looking tables that they have the appearance of being stratified. The slaty masses are, generally, vertical; but in one hill they were observed dipping 80° to the south-east. The direction of the tabular masses is mostly across the oblong summits of the hills. The appearances of stratification were not observed to extend through a whole hill, and seemed, in fact, to be confined to the more decomposable granites; but the naked rocks are every where traversed by smooth fissures. The blocks, which lie under the cliffs, have sometimes a tabular form, but more generally come nearer to a cube or rhomboid, and present one or two very even faces. Few veins were noticed. In the more sheltered vallies, some clumps of white or black spruce trees occur; but the hills are barren.

The point of land which lies between M'Tavish and M'Vicar Bays has low shores; but five or six miles inland, an even-backed ridge rises gradually to the height of three or four hundred feet, and abuts obliquely against the primitive hills. I did not visit this ridge, and the snow prevented me from seeing any flat beds of rocks, if such exist on the shore. On one point, however, near the north end of Dease Bay, many large angular blocks of whitish dolomite were piled up, and I have little doubt of the rock existing *in situ* in that immediate neighbourhood.

M'Tavish Bay is forty miles long, and twenty wide, and its depth of water, near the eastern shore, exceeds forty-five fathoms. Some shoals of boulders skirt the coast near Point Leith. M'Vicar Bay is about seventy miles long, and from eight to twelve wide; and at the "fishery," in a narrow part, not far from its bottom, its depth of water, two miles from the shore, is twelve fathoms. Dease Bay is equal to M'Tavish Bay in extent, and opens to the S.W. into the body of the lake. The high lands at the N.E. end, or bottom of this bay, have an even outline, and appear to attain an elevation of eight or nine hundred feet, at the distance of six or seven miles from the shore. Near its east side lie the lofty islands of Narrakazzæ, which rise seven hundred feet above the lake. Dease River, the principal feeder of the lake, falls into the bottom of Dease Bay. It is two hundred yards wide, and from one to three fathoms deep near its mouth. A few miles up this river a formation of soft red sandstone occurs, which will be noticed hereafter.

LIMESTONE.—GREAT BEAR LAKE.

At the mouth of Dease river there are hills five or six hundred feet high, composed principally, or entirely, of 228* dolomite in horizontal strata. Some of the beds consist of a thick-slaty, fine-grained dolomite, containing dispersed scales of mica, which is most abundant on the surfaces of the slates. Most of the beds, however, consist 228 of a thin-slaty, dull, purplish dolomite, traversed by veins of calc-spar. The structure of this rock is compact, approaching to fine granular; and some of the beds have what quarry-men term "clay-facings," that is, they are encrusted with a thin film of indurated clay.

Greenstone slate? occurs in horizontal beds on the north shore, eight or nine miles to the westward of Dease River:

and at Limestone Point,* about twenty miles from the river, a small range of hills terminates on the borders of the lake, in shelving, broken cliffs, about two hundred feet high. These cliffs consist chiefly of nearly compact light-coloured dolomite, interstratified with greenstone, and a brownish-red limestone, such as occurs in the hills at the mouth of the Dease River. In contact with the greenstone, there is a bed of talcose limestone, having a curved, slaty structure: most of the beds of dolomite are hard, and pass into chert.

ALUMINOUS SHALE.—GREAT BEAR LAKE.

The north shore of Bear Lake is low, and is skirted by many shoals, formed by boulders of limestone. No rocks, *in situ*, are exposed between Limestone Point and the Scented Grass Hill, a remarkable promontory, which separates Smith and Keith bays. Its height above the lake is betwixt eight and nine hundred feet, and in form and altitude it corresponds with the Great Bear Mountain, which, lying opposite to it, sepa-

* Section of the cliffs at Limestone Point—strata dipping to the N.N.W.

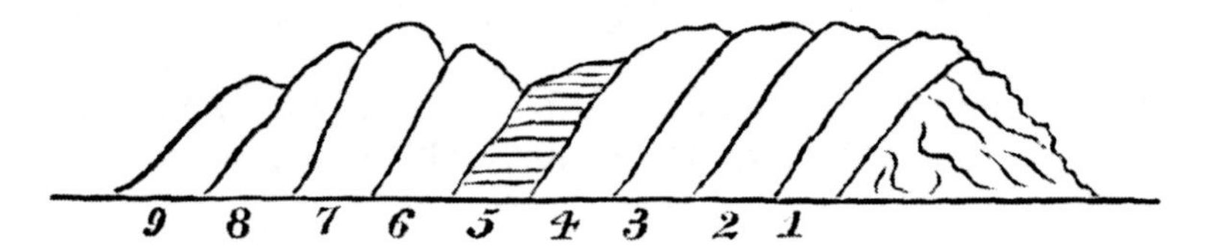

In the section the strata are represented much more inclined than they really are.

231 Fine-grained, nearly compact, yellowish-gray dolomite, forming the summit of the hill, but the first, or lowest stratum, in the language of geologists.

232 Compact, splintery dolomite, with a conchoidal fracture, and wax-yellow colour—second stratum.

233 A cherty dolomite; containing calc-spar—third stratum.

234 Bluish-gray dolomite, traversed by calc-spar—is nearly compact, and has an uneven, splintery fracture—forms the uppermost portion of the fourth stratum.

235 Talcose? limestone, having a curved slaty structure, and containing cherty portions—from the lower part of the fourth stratum.

236, 237 Earthy greenstone? forms the fifth stratum.

238 Brownish-red dolomite, with an uneven fracture, scarcely splintery. It has a compact structure, and is intersected by veins of calc-spar—from the sixth stratum.

239 Light yellowish gray dolomite, passing into chert—seventh stratum.

240, 241 Thin slaty beds of brownish-red dolomite, like 238—eighth stratum.

242 Bluish-white porcelain chert, sometimes mixed with red dolomite—243 —ninth stratum.

rates M'Vicar and Keith bays. I did not ascend either of these hills; but cliffs, corresponding in character to those of the aluminous shale-banks at Whitby, flank their bases; and the same formation probably extends along the north shore of Keith Bay, and some way down Bear Lake River. The ground skirting the Scented Grass and Great Bear Mountains is much broken, and consists of small, rounded and steep eminences, separated by narrow vallies and small lakes. Several shelving cliffs, about one hundred feet high, and some miles in extent are washed by Bear Lake. They consist of slate-clay and shale, more or less bituminous, and the dip of the
251 strata is in several places to the N.W. by N. At the foot of the Scented Grass Hill a rivulet has made a section to the depth of one hundred feet, and here the shaly
244 beds are interstratified with thin layers of blackish-brown,
246 earthy-looking swinestone, containing selenite and py-
247 rites. Globular concretions of the same stone, and of a poor clay iron-stone, also occur in beds in the shale. The surfaces of the slates were covered with an efflorescence
249 of alum and sulphur. Many crystals of sulphat of iron
250 lie at the bottom of the cliff, and several layers of plu-
248 mose alum, half an inch thick, occur in the strata. At the base of Great Bear Mountain, the bituminous shale is interstratified with slate-clay, and I found imbedded in the former a single piece of brown coal, in which the fibrous structure of wood is apparent. Sections of slate-clay banks, and more rarely of bituminous shale, occur in several places on the north shore of Keith Bay. In one place, about seven or eight miles from Bear Lake River, a bed of plastic and bituminous clay occurs, and in another, near Fort Franklin, there is a deposit of an earthy coal, which possesses the characters of *black chalk*.

It is probable that a magnesian limestone underlies this formation of bituminous shale. I have already mentioned the beds of dolomite, which are exposed on the north side of Bear Lake, and similar beds occur to the southward of the Great Bear Mountain, forming cliffs on the shores of M'Vicar Bay. At Manito Point, on the west side of the isthmus that connects Great Bear Mountain to the main shore, a low ridge of limestone rocks terminates on the borders of the lake, forming some bold cliffs and a remarkable cave. The stone has a gray colour and bituminous smell, and contains much interspersed calc-spar. The strata dip to the north-west.

VICINITY OF FORT FRANKLIN, GREAT BEAR LAKE.

Fort Franklin stands on the northern shore of Keith Bay, about four miles from Bear Lake River, upon a small terrace, which is elevated twenty-five or thirty feet above the lake. The bay, contracting towards the river, is about four miles wide opposite to the fort, and the depth of water there does not exceed four fathoms. Farther from the river, the east and west shores of Keith Bay recede to the distance of thirty miles from each other, and the depth of water in the centre of the channel greatly increases. The bottom of this bay, wherever it could be distinguished, was observed to be sandy, and thickly strewed with round boulders* of various primitive rocks of

* *List of boulders gathered on the beach at Fort Franklin.*

261 Coarse crystalline granite; felspar flesh-red in large crystals; quartz gray; mica black.

262 Granite; felspar paler, and less distinctly crystallized; quartz in small quantity, gray; mica blackish, and rather abundant.

263 Granite; felspar partly reddish, partly yellowish-white, quartz in small grains; mica equalling the quartz in quantity, black.

264 Granite, fine-grained; quartz and felspar, white, the former nearly transparent, black mica in small specks, garnets.

265, 268 Granite; quartz in regular crystals; mica blackish, in small quantity.

266 Granite? red felspar in large crystals; quartz gray; mica replaced by chlorite?

267 Granite; felspar gray; chlorite? in small quantity.

269 Granite, small grained, passing into gneiss; reddish-brown felspar and gray quartz, intimately mixed, and having in the aggregate, a vitreous lustre; mica in layers.

270 Granite coarser grained than the preceding, containing more quartz; the mica disseminated.

271, 273 Granite with little mica, some portions of the felspar tinged green.

272, 274, 275, 277, 278, 279, 280, } Granite grayish and small grained mica black.

276 Granite; brick-red felspar; quartz; and augite?—no mica.

The mica is mostly black in all the granite boulders that occur here, the felspar most frequently reddish.

281 Porphyritic granite? felspar imperfectly crystallized, containing large, imbedded crystals; quartz; and chlorite?

282 Granite? composed of felspar, of quartz, with, perhaps, a few minute grains of chlorite?

283 Granite? contains little quartz, and a few scales of mica, with some chlorite?

284 Sienite; felspar somewhat granular, a little quartz and chlorite?

285 Porphyritic sienite? having a basis of slightly granular felspar, with light-coloured crystals of felspar, some quartz and disseminated grains of chlorite?

286 Reddish-brown hornstone porphyry.

287 Crystalline greenstone.

288 Fine-grained greenstone.

289 Porphyritic greenstone.

290 Pitchstone porphyry.

291 Greenstone slate with pyrites.

large size, which were particularly abundant near the river, and with large square blocks of limestone, most plentiful near the cape formed by the Scented Grass Hill. In the small bay between the fort and the river, shoals are formed by accumulations of boulders, and the shores are thickly strewed with them. Many of these travelled blocks consist of flesh-red granite, having only a small quantity of black mica, exactly resembling the primitive rocks seen in M'Tavish Bay, but noticed no where else near the lake. Boulders of the same description occur in shoals at the mouth of M'Tavish Bay, and on the shores which skirt the Scented Grass Hill which faces that bay, to all which places they may have been brought from the parent rock, by a current flowing from the east. On the northern shore of Bear Lake the great majority of the boulders consists of limestone. Two varieties of granite, which occur amongst the boulders, were recognised as being abundant rocks at Fort Enterprise, which is situated about one hundred and seventy miles south-east from M'Tavish Bay. Some of the boulders were of a peculiar-looking porphyry exactly resembling that which occurs in the height of land betwixt the Coppermine River and Dease Bay; several of sandstone and conglomerate, which probably came from the same quarter; of greenstone, perhaps, from the Copper Mountains, and of limestone from the northern shores of the lake, and from the isthmus of the Great Bear Mountain; all these places lying to the eastward or north-east.

261 to 308

266 282

The soil in the immediate vicinity of Fort Franklin is sandy, or gravelly, and covers, to the depth of one or two feet, a bed of clay of unknown thickness. Gravel taken from a spot thirty feet above the present high-water level of the lake, and out of the reach of any stream or torrent, contained rounded pebbles of granite, of greenstone, of quartz rock, of lydian stone, and of various sandstones, of which some were spotted, and

292 Amygdaloidal claystone porphyry.
293 Compact grayish-blue dolomite.
294 Splintery dolomite.
295 Cellular dolomite.
296 Swinestone.
297 Limestone with corallines.
298 Chert.
299 White quartz.
300 Quartz-rock.
301 Coarse sandstone.
302 Fine-grained white sandstone.
303 Fine-grained red sandstone;
304 Fine-grained striped sandstone.
305 Fine-grained spotted sandstone.
306 Slaty sandstone verging towards slate-clay.
307 Dark-red claystone.
308 Light-coloured claystone.

others presented zones of different colours. These sandstones form a considerable portion of the gravel.*

The clay which lies under the soil is of a bluish-gray colour, and is plastic but not very tenacious. It is more or less mixed with gravel. During the greater part of the year it is firmly frozen; the thaw in the two seasons we remained there never penetrating more than twenty-one inches from the surface of the earth. In spots where the sandy soil is wanting, the clay is covered a foot deep, or more, by moses, mostly *bryum palustre*, and some marsh *hypna* and *dicrana*, in a living state, for they seem to be converted very slowly into peat in this climate.

The ground rises gradually behind the fort, until it attains, at the distance of half a mile from the lake, the height of two hundred feet, forming, when viewed from the southward, an even ridge, running nearly east and west—which ridge is, in fact, the high bank of the lake, as it corresponds in height with the summit level of the banks of Bear Lake River, and of the southern shore of Keith Bay. The country extending to the northward, from the top of the bank, is nearly level, or has a very gentle ascent for about five miles, when a more abrupt ridge rises to perhaps three hundred or four hundred feet above the lake. The view from the summit of this second eminence is very extensive, the whole country as far as the eye can reach appearing to be a level, from which several

* *List of Specimens from Diluvial Gravel, Fort Franklin.*

1 Amphibolic granite, rather coarse crystalline, felspar flesh-red.
2 Ditto, approaching to gneiss.
3 Gneiss approaching to mica-slate, felspar white, and in small quantity.
4 Greenstone with much felspar and minute disseminated pyrites.
5 Quartz rock? having brownish and imperfect crystals, and a reddish disintegrated mineral disseminated.
6 Brownish-red and fine granular quartz-rock, with a somewhat splintery fracture. It has the aspect of compact felspar.
7 Quartz rock, reddish crystalline texture, and vitreous lustre, but with small rounded grains imbedded in it, bringing it near to sandstone.
8 Coarse sandstone; rounded grains of quartz united by a clayey basis.
9 Fine-grained purplish sandstone, with grayish spots. This sandstone occurs *in situ* near the Copper Mountains, between Dease Bay and the Coppermine River.
10 Fine-grained yellowish-white sandstone.
11 Yellowish-gray sandstone, composed of small rounded grains of quartz united by a powdery white basis.
12 Yellowish-gray sandstone, composed of fine grains of vitreous quartz.
13 Sandstone, having different shades of brownish-red colour, in layers.
14 Lydian stone.

narrow precipitous ridges of limestone arise. But, although the country around these ridges appears from a distance to be level, or very slightly undulated, yet it abounds in small eminences and steep-sided vallies of various shapes, some being rounded and basin-shaped, others long and narrow. Lakes and swamps are here so numerous, that the country, for at least sixty miles to the northward, is impassable in summer, even to the natives. There are many mounds of sand and gravel, and fragments of sandstone are frequent; but having travelled in this direction only in winter, when the ground was covered to the depth of upwards of three feet with snow, I had not an opportunity of examining its geological structure. White spruces cover the drier spots; larches, black spruces, and willows abound in moist places; the sandy hillocks are clothed with aspens, and the sides of the vallies support some canoe birches, with a thick undergrowth of dwarf birches, alders, and rose-bushes. The eminence from whence the view just described was obtained, appears like a ridge only in approaching it from the lake, for it rises very little above the general level of the country behind it. It has a direction from N.W. by N. to S.E. by S., and terminates about eight miles to the eastward of the fort, in a small bluff point on the shores of the lake and there the strata consist of slate-clay slightly bituminous. The banks immediately behind the fort also exhibit, in their ravines, a bluish slate-clay.

The land on the south side, or bottom, of Keith Bay, presents a nearly similar aspect to that just described, rising, on the borders of the lake, to the height of one hundred and fifty feet, and then running back to a great distance nearly level. It may be characterized as full of hollows, narrow vallies, ravines, and lakes; but it is not hilly, although it is traversed by ridges of limestone, which rise like walls through the flat country. The nearest of these ridges terminates on the borders of the lake at the *Manito Point*, (noticed in page vii.) It may be proper to remark here, that, in addition to the limestone ridges visible from Fort Franklin, or from the heights behind it, the summit of Clark Hill, bearing south, and forming part of a ridge about fifty miles distant, was distinctly seen. This hill lies behind Old Fort Norman on the Mackenzie, and has more the outline of a granitic rock, although some of the peaks which skirt it have the serrated crests which the limestone ridges in this quarter show. It was guessed to be 1500 feet high above the Mackenzie.

This sketch of the general features of the country about

Fort Franklin being premised, the ensuing geological notices follow in the order of the route of the Expedition.

BEAR LAKE RIVER—SANDSTONE, LIMESTONE.

Bear Lake River is about seventy miles long, from its origin in the lake till it falls into the Mackenzie, and throughout its whole length, its breadth is never less than one hundred and fifty yards, except at the *Rapid*, a remarkable place, about the middle of its course. It is from one to three fathoms deep, and very rapid, its velocity being estimated at six miles in the hour. Its waters are clear as they issue from the lake, but several branches of considerable size bring down muddy water, particularly one which flows from the north, and falls in below the rapid.

Above the rapid, the valley of the river is very narrow, the banks every where sloping steeply from the level of the country. Their summit line, which is nearly straight, is about one hundred and fifty feet above the bed of the river. In some places they have an even face elevated at an angle of about forty-five degrees, and they are not unfrequently cut by ravines into pretty regular figures, resembling hay-ricks, or the parapet of a fort, the ravines representing the embrasures. Sections made by the river presented generally sand or clay; the sand probably proceeding from the disintegration of a friable, gray sandstone, which showed itself occasionally in a more solid form. The rapidity of our voyage, however, afforded us little opportunity of searching for the solid strata which are generally hid by the debris of the bank. About twelve miles above the rapid, a small-grained, friable sandstone, of a yellowish gray colour, and irregular earthy fracture, is associated with beds of bluish-gray slate-clay. These beds consist of concretions of various sizes and irregular shapes, but which may be said to approach in general to a depressed orbicular form; their surfaces are coloured purplish-brown by iron, and studded with crystals of sulphate of lime. This slate-clay contains many small round grains of quartz, and is exactly similar to that which occurs at the rapid, and which will be afterwards noticed. In other places the banks are covered by the debris of a slate-clay slightly bituminous, resembling wacke in its mode of disintegrating.

The *Rapid* is caused by the river struggling through a chasm bounded by two perpendicular walls of sandstone, over an uneven bed of the same material. On escaping from this narrow passage, it winds round the end of a lofty cliff of limestone, which forms part of a ridge that is continued through the country on both sides of the river.

Viewed from the summit of this ridge, which rises about eight hundred feet above the river, the country towards Bear Lake appears level. The view down the river presents also a plain country, bounded on the Mackenzie by another limestone ridge, which, unless the eye was deceived by the distance, gradually inclined to the one at the rapid, and appeared, by joining it to the northward, to form a great basin. These ridges are also prolonged to the southward. The plain is covered with wood, intersected by chains of lakes, and seemed to lie rather below the summit level of the banks of Bear Lake River. It is only comparatively, that the country deserves the name of plain, for its surface is much varied by depressions, ravines, and small eminences, that do not, however, destroy the general level appearance when seen from a distance. The view from the hill is terminated, to the westward, by the distant chain of the Rocky Mountains, running nearly N.W. by N. A little below the rapid, a small stream from the southward flows into the Bear Lake River, near whose sources the Indians procure an excellent common salt, which is deposited from the springs by spontaneous evaporation.

The walls of the rapid are about three miles long, and 120 feet high. They are composed of horizontal beds, the lower of which consist of an earthy-looking stone, intermediate
25 between slate-clay and sandstone, having interiorly a dull yellowish-gray colour. Concretions, with smooth surfaces, about the thickness of a swan's quill, pass perpendicularly through the beds like pins, are prolonged beyond the partings, and bear some resemblance to portions of the roots or branches of a tree. The seam surfaces are very uneven. These beds are parted by thin slaty layers, of a stone similar in appearance, but rather harder, and containing many interspersed scales of mica, and also some minute portions
18 of carbonaceous matter in the form of lignite. The thin layers contain impressions of ferns, and from the
19 debris at the bottom off the cliff I gathered impressions
1827 of the bark of a tree (lepidodendron) and some am-

monites in a brown iron-shot sandstone.* The upper beds are composed of a fine grained, quartzose, gray sandstone, having an earthy basis, and occasionally interspersed carbonaceous matter. Some of the beds are a foot and a half thick, and have sufficient tenacity to be fitted for making grindstones; most of the sandstone is, however, rather friable. Near the summit there is interposed a bed of fine-grained dolomite, and a friable sandstone, which forms the crest of the cliff, and exhibits in its weathering battlement-shaped projections and pinnacles. Covering this sandstone, but not quite to the margin of the cliff, there is a layer of slaty limestone, having a bluish or blackish-gray colour, a dull fracture, and rather compact structure. In the lower beds of the cliff there are some globular and disk-shaped concretions, of an indurated iron-shot slate-clay, or poor clay-iron-stone, containing pyrites. They vary in magnitude from six inches to a foot and a half in diameter, and appear to be formed of concentric layers, which are rendered apparent by the weathering of the stone. The sandstones and shales of the rapid have a strong resemblance in appearance to those of the coal measures; but pitch-coal was not detected at this place. Several distinct concretions of indurated slate-clay, assuming the appearance termed *cone in cone*, were picked up among the boulders on the banks of Bear Lake River, some way below the rapid, but they were not traced to their parent beds. They effervesce with acids.

18,20,21, 22,23,24, 26,27,

28

29

30

Between the walls of the rapid and the limestone ridge there is a piece of meadow-ground, having a soft, clayey soil, in which, near the base of the hill, a small rivulet flows to join the river. The bed of this rivulet presents accumulations of boulders of large size, arranged so as to form two terraces, the upper of which is considerably above the highest level either of the rivulet, or of Bear Lake River. The boulders consist of varieties of granite, gneiss, mica-slate with garnets, greenstone and porphyry. One of the porphyries is a beautiful stone, composed of hyacinth-red felspar, and irregular crystals of milky quartz, with a few specks of a dark

50

* Mr. Sowerby, who inspected all the specimens containing organic remains, says of this species of ammonite, "it is, as far as I can discover, new. It contains sulphate of barytes, and is probably referrible to some of the Oolites near the Oxford clay." Although it was found lying on the beach, I have no doubt of its having fallen from some of the beds of clayey sandstone, which form the walls of the rapid.

green mineral, and very much resembles a rock which is not uncommon in the gneiss districts about Fort Enterprize.

45 Many of the boulders consist of conglomerates and sand-
47 stones that strongly resemble those of the old red sand-
50 stone formation, which forms the height of land between
51 Dease Bay and Coppermine rivers. Also some flinty
49 slates, mixed, in thin layers, with compact, yellowish limestone, and some pebbles of jasper interleaved with flinty slate.

The limestone ridge below the rapid stands on a narrow base, whose transverse diameter does not exceed a quarter of a mile. Its summits are generally conical, but very rugged and craggy; the highest peak I had an opportunity of visiting is about a mile from Bear Lake River, and it has been already stated to be estimated at eight hundred feet above that stream, or nine hundred and fifty above the sea. The general direction of the ridge is from S.E. by S. to N.W. by N., or nearly parallel to the great Rocky Mountain chain, and to the smaller ridges betwixt it and that chain. Its prolongation through the flat surrounding strata, to the southward of Bear Lake River, can be traced for at least forty miles, and it is visible at nearly an equal distance, as it runs through the still more level country to the northward; but here, as has been already said, it appears to incline towards the similar ridge which is cut by the Mackenzie, at the mouth of the Bear Lake River, and is about twenty-five miles to the W.S.W., in a direct line. That part of the ridge which I had an opportunity of visiting, consisted entirely of limestone, generally in thick beds. Its stratification was not very evident, and in my very cursory examination the general dip was not clearly ascertained. A precipitous cliff, four hundred feet high, facing the S.E., and washed by the Bear Lake River, presents strata, inclined to the S.W. at an angle of 45°, which may be perhaps considered as the general dip; for the ridge on that side slopes down to the surrounding country at an angle of about 30° or 40°, while on the N.E. side it presents lofty precipices formed by the cropping out of the strata. Many of the beds in this hill consisted of a blackish-gray fine grained limestone, intersected by veins
39,34 of calc-spar; but several layers of gray and dark
40 coloured dolomites, and some of a yellowish-gray *rauchwacke*, were interstratified with them, and the upper parts of the precipitous cliff, and also of the highest peak, consisted of a calcareous breccia, containing
35,36 rounded pieces of brown limestone, and angular frag-

ments of chert; and the faces of some cliffs, on the 42,43,44 N.E. side of the hill, were incrusted with a fine crystalline gypsum to the depth of from one to two feet.*

The banks of Bear Lake River below the rapid have a more gentle declivity than those above it, and they occasionally recede from the stream, so as to leave a grassy slope varying from a few yards to half a mile in breadth. The sections of these banks by torrents present only sand or clay; and the hollows of the ravines are lined with boulders principally of primitive rocks. No stone was observed *in situ* from the rapid until we came to the junction of the river with the Mackenzie.

The Bear Lake River flows into the Mackenzie at a right angle, and on its north bank, at its mouth, there is a hill, which has been already noticed as forming part of a ridge visible from the one at the rapid, with which it probably unites to form a great basin. These two hills seem to belong to the same formation. The body of the hill consists of highly-61,62 inclined beds of blackish-gray limestone, with sparry 60 veins, and of brownish-gray dolomite, which cannot be distinguished in hand specimens from that of the hill at the rapid. The superior beds are formed of a calcareous

* 33 This limestone appears as if composed of an aggregate of small crystals, and presents many drusy cavities.

34 Is an adjoining bed of a similar colour, of a fine crystalline texture, but without the drusy cavities. It appears to be a dolomite. These two beds dip to the northward.

35, 36 Calcareous breccia. The two preceding beds (33 and 34) were from the summit of the portion of the hill which forms the cliff, but taken a little farther to the N.W. In the cliff the beds dip, as has been stated, to the S.W. The following beds occur in going to the north-westward, towards the summit of the highest peak, commencing near its base, in a valley behind the cliff.

37 A fine-grained blackish-gray dolomite, having interspersed many nodules of chert, or grayish-white quartz, not crystallized.

38 A very compact, opaque limestone, of a smoke-gray colour, having a flat and slightly splintery fracture. Effervesces briskly.

39 Blackish-gray rather compact limestone, having a flat and dull fracture, and intersected by small veins of calc-spa. This is a prevalent stone in the hill, and also occurs in quantity in other limestone ridges in the neighbourhood.

40 An ash-gray, fine-granular dolomite.

41 A conglomerate, forming the summit of the highest peak.

57 breccia.* Associated with these strata, however, there
58,59 are beds of limestone, highly charged with bitumen;
63,64 and at the base of the hill there are beds of bituminous
65 shale, some of which effervesce with acids, whilst others approach in hardness, and other characters, to flinty slate. These shaly beds were seen by Captain Franklin and Mr. Kendall in autumn 1825, and they also saw, at that time, some sulphureous springs and streams of mineral pitch issuing from the lower parts of the limestone strata; but the whole of them were hid by the height of the waters of the Mackenzie in the spring of 1826.† The same cause prevented
69,66 me from seeing some beds of lignite and sandstone, at
67,68 the same place, of which Captain Franklin obtained specimens.

LIGNITE FORMATION.—MACKENZIE'S RIVER.

Having noticed the general features of this portion of the river, I have next to state, that the formation constituting its banks may be characterized as consisting of wood-coal in va-

* 57 This breccia has a white calcareous basis, which incloses angular fragments of compact, yellowish-gray limestone, with smooth dull surfaces.

58 Grayish-white limestone, having a fine crystalline texture, with drusy cavities, incrusted with bitumen.

59 Limestone, apparently composed of crystalline fragments, highly charged with bitumen, cemented by a whitish carbonate of lime in minute crystals. I could not satisfy myself whether this variety of colour proceeded from partial impregnations of bitumen, or from a brecciated structure. Specimens 58 and 59 were from beds near the western part of the hill.

60 A fine-grained dolomite, approaching to compact, having a flat and somewhat splintery fracture, and a brownish-gray colour.

61, 62 Limestone in the body of the hill, resembling No. 39 in the hill at the rapid in Bear Lake River, but with larger veins of calc-spar.

63, 64 Dark blackish-brown bituminous shale, veined with calc-spar, and passing into bituminous marl-slate. It contains nodules of iron pyrites.

65 Thin bed of indurated shale, approaching to flinty-slate, lying at the foot of some beds of bituminous limestone. Their connection not clearly made out.

66, 67, 68 Bluish-gray, fine-grained sandsone, some of them passing into slate-clay, and scarcely to be distinguished from those at the rapid in Bear Lake River. Capt. Franklin took these specimens from horizontal beds at the foot of the hill facing Bear Lake River.

† Sir Alexander Mackenzie, in p. 95 of his Voyage to the Arctic Sea, states, that he saw several small mineral springs running from the foot of this mountain, and found lumps of iron ore on the beach.

rious states, alternating with beds of pipe-clay, potter's clay, which is sometimes bituminous, slate-clay, gravel, sand, and friable sand-stones, and oscasionally with porcelain earth. The strata are generally horizontal, and as many as four beds of lignite are exposed in some parts, the upper of which are above the level of the highest river-floods of the present day.

The *lignite*, when recently detached from the beds, is pretty compact, but soon splits into rhomboidal pieces, which again separate into slates more or less fine. It burns with a very fetid smell, somewhat resembling that of phosphorus, with little smoke or flame, leaving a brownish-red ash, not one-tenth of the original bulk of the coal. The blacksmith found it unfit for welding iron when used alone, but it answered when mixed with charcoal, although the stench it created was a great annoyance. Different beds, and even different parts of the same bed, presented specimens of the fibrous brown-coal, earth-coal, conchoidal brown-coal, and trapezoidal brown-coal of Jameson. Some of the pieces have the external appearance of compact bitumen, but they generally exhibit, in the cross fracture, the fibrous structure of wood in concentric layers, apparently much compressed. Other specimens have a strong external resemblance to charcoal in structure, colour, and lustre. A frequent form of the lignite is that of slate, of a dull, brownish-black colour, but yielding a shining streak. The slate is composed of fragments, resembling charred wood, united together by a paste of more comminuted woody matter, mixed, perhaps, with a small portion of clay. In the paste there are some transparent crystals of sulphate of lime, and occasionally some minute portions of a substance like resin. These shaly beds bear a strong resemblance to peat, not only in structure but also in the mode of burning, and in the light whitish ashes which are left. The external shape of stems or branches of trees, is best preserved in some fragments impregnated with slate-clay, and occasionally with siliceous matter, which occur imbedded in the coal. The bark of these pieces has been converted into lignite. Some of them exhibit knots, such as occur where a branch has decayed, and others represent the twists and contortions of wood of stunted growth. The lignite is generally penetrated by fibrous roots, probably *rhizomorpha*, which insinuate their ramifications into every crevice.

The beds of lignite appear to take fire spontaneously when exposed to the atmosphere. They were burning when Sir Alexander Mackenzie passed down the river in 1789, and have been on fire, in some part or other of the formation, ever since. In consequence of the destruction of the coal, large slips of the bank take place, and it is only where the debris has been washed away by the river that good sections are visible. The beds were on fire when we visited them, and the burnt clays, vitrified sand, agglutinated gravel, &c. gave many spots the appearance of an old brick-field.

81 The *gravel* interstratified with the lignite, consists of smooth pebbles of Lydian stone, of flinty slate, of white quartz, of quartzose sandstone, and conglomerate, like the sandstones and conglomerates of the old red sandstone formation, of clay-stone, and of slate-clay, varying in size from a pea to that of an orange. The gravel is often intermixed with a little clay, which gives the bed sufficient tenacity to form cliffs, but does not prevent the pebbles from separating, in the attempt to break off hand specimens. It is seamed by thin layers of fine sand: beds of sandstone are of occasional occurrence.

Potter's clay occurs in thick beds, has generally a gray or brown colour, and passes, in some places, into a highly bituminous thick-slaty clay, penetrated by ramifications of carbonaceous matter resembling the roots of vegetables.

The *pipe-clay* is deserving of particular notice. It is found in beds from six inches to a foot thick, and mostly in contact with the lignite. It has commonly a yellowish-white colour, but in some places its hue is light lake-red. The natives use it as an article of food in times of scarcity, and it is said to have sustained life for a considerable time. It is termed *white mud* by the traders, who whitewash their houses with it. It occurs also in lignite deposits on the upper branches of the Saskatchewan, and is associated with bituminous shale on the coast of the Arctic Sea. Mr. Nuttall mentions a similar substance, under the name of pink-clay, as being found in the lignite deposits on the Arkansa.*

The *porcelain earth* was observed only at one place where the beds were highly inclined, and there it appeared to replace the sandstones of other parts of the deposit. It has a whitish colour, and the appearance, at first sight, of chalk; but some of its beds, from the quantity of carbonaceous matter interspersed through them, having a grayish hue. Its beds are from two to three yards thick.

* Travels in the Arkansa, p. 52—54.

In a note* I have mentioned the most remarkable sections of this formation which occur on the banks of the Mackenzie.

* Section I.

The section of the bank at the mouth of the Bear Lake River is as follows, beginning with the lowest bed:—

81	Gravel, with thin layers of sand rising from the water's edge in a perpendicular cliff, to the height of - -	30 feet
	Lignite (70 to 80 and 84) - - - - - - -	1
83	Potter's clay of a bluish gray colour, alternating with layers of sand - - - - - - -	40
	A sloping uneven brow, covered with soil, extends to the summit of the bank - - - - - -	20
		91

Lydian stone is the most abundant, and whitish quartz the least so of the pebbles mentioned in the text as entering into the composition of the gravel.

A little farther up the Mackenzie, this bed of gravel passes into sand, 82 which, in some spots, has sufficient coherence to merit for it the name of sandstone. During a great part even of the summer season, all the beds of sand are frozen into a hard sandstone; but a piece having been broken off and put into the pocket, speedily thawed into sand.

83 Specimens of the clay, which I have denominated potter's clay, taken from near the beds of lignite, have a colour intermediate between yellowish-gray and clove-brown, a dull earthy fracture, and a slightly greasy feel. It is not gritty under the knife, and acquires a slightly shining smooth surface, adheres slightly to the tongue, and, when moistened with water, assumes a darker colour, and becomes plastic.

Section II.

About five miles above Bear Lake River, the cliff consists of Slaty sandstone evidently composed of the same materials with the friable kinds described in the text, but having tenacity enough to form a building stone. It incloses some seams of lignite - - - - - - -	10 feet
Lignite - - - - - - - - - - -	4½
Clay and Sand - - - - - - - - -	50
Irregular slope from top of cliff to summit of bank - -	90
	154½

Section III.

A little farther up the river than the preceding:—

85	Pipe-clay on a level with the water - - - -	1 foot
86	Lignite - - - - - - - - - -	1
90	Potter's clay - - - - - - - - -	14 feet
87	Pipe-clay - - - - - - - - - -	1 foot
89	Lignite - - - - - - - - - -	1
91	Potter's clay - - - - - - - - -	10 feet
	Lignite - - - - - - - - - -	1 foot
	Sandstone - - - - - - - - - -	8 feet
	Lignite - - - - - - - - - -	2½
	Potter's clay - - - - - - - - -	10

The depth of the formation was not ascertained, but the sections will show the thickness of the beds which were exposed.

94	Friable sandstone and clay - - - - - -	20 feet
	Sandstone a little more durable - - - - -	12
	Sloping Summit - - - - - - - -	40
		121½

The pipe-clay, when taken newly from the bed, is soft and plastic, has little grittiness, and when chewed for a little time, a somewhat unctuous but not unpleasant taste. When dried in the air it acquires the hardness of chalk, adheres to the tongue, and has the appearance of the whiter kinds of English pipe-clay, but is more meagre.

Section IV.

A little above the preceding:—

A precipitous bank of gravel - - - - - -	12 feet
Lignite and clay, the beds concealed by debris -	40
Friable sandstone - - - - - - - -	30
Height of the cliff - - - -	82

Section V.

Ten miles above Bear Lake River, at the junction of a small torrent with the Mackenzie, there is a cliff about forty feet high, in which the strata have a dip of sixty degrees to the southward.

	Bed, No.		
98	1	Porcelain clay - - - - - - -	2 yards
	2	Potter's clay slightly bituminous -	1
99, 100, 101	3	Thin-slaty lignite, with two seams of clay-iron stone, an inch thick - - - - -	2½
	4	Pipe clay, (nine inches) - - - -	¼
104	5	Porcelain clay - - - - - - -	3
105	6	Bituminous clay - - - - - -	3
106	7	Lignite, with a conchoidal fracture -	2
	8	Pipe clay - - - - - - - -	¼
107	9	Porcelain clay - - - - - - -	3
	10	Bituminous clay - - - - - -	3
110	11	Lignite, earthy paste, enclosing fibrous fragments - - - - - - - -	2
	12	Porcelain earth - - - - - - }	
	13	Bituminous clay - - - - - - }	9
	14	Porcelain earth - - - - - - }	
			31 yards.

The three last beds it is probable, once inclosed seams of coal which have been consumed, but the quantity of debris prevented this from being ascertained satisfactorily during the hurried visit I paid to them.

Over these inclined beds there is a shelving and crumbling cliff of 108 sand and clay covered by a sloping bank of vegetable earth. A layer of peat at the summit has a thin slaty structure, and presents altogether, except in colour and lustre, a striking resemblance to the shaly lignite, forming bed No. 3 in the preceding Section.

The height above the sea of the summit of the banks it forms on the Mackenzie, was estimated to be from two hundred and fifty to three hundred feet.

NOTICES OF OTHER LIGNITE FORMATIONS.

Similar formations of lignite occur near the foot of the Rocky Mountain range farther to the southward; but I have not, after many inquiries, heard of any traces of them in the eastern parts of the Hudson's Bay lands. Sir Alexander Mackenzie, after describing the general course of the Rocky Mountains, says that "along their eastern edge, there occurs a narrow strip of marshy, boggy, and uneven ground, which produces coal and bitumen;" and that "he saw these on the banks of the Mackenzie in lat. 66°, and, in his second journey, on the Peace River, in lat. 56° and 146° W. long;" and further, that "the same was observed by Mr. Fidler, on the south branch of the Saskatchewan, in lat. 52° long. 112½° W." Mr. Alexander Stewart, an intelligent chief factor of the Hudson's Bay Company, and well acquainted with those countries, informs me that there are beds of coal on fire, on the Smoking River, or east branch of the Peace River, and on the upper parts of the *Rivière la biche*, or Elk River; and that coal, although not on fire, occurs at Lesser Slave Lake, on a line with the other two localities. Mr. Small, a clerk to the Hudson's Bay Company, likewise acquaints me, that coal occurs at Edmonton, on the north branch of the Saskatchewan, in beds, sometimes seven or eight feet thick. Most of the coal is thin-slaty; but some beds yield shining, thick lumps, which break, as he expresses it, like Spanish liquorice. It lies over beds of bluish-gray sandstone, and is associated with a white clay, which froths in water and adheres to the fingers.

104, 98. The substance composing beds Nos. 1 and 5, which I have denominated Porcelain clay, has a fine, granular texture, and the appearance of some varieties of chalk. It adheres slightly to the tongue, yields readily to the nail, is meagre, and soils the fingers slightly. There are many specks of coaly matter disseminated through it, and some minute scales of mica, and perhaps of quartz. When moistened with water, it becomes more friable, and is not plastic. It does not effervesce with acids.

Bed No. 9 is the same mineral that forms beds 1 and 5; but it has a grayer colour from the greater quantity of coaly particles, and its structure is slightly slaty.

The bituminous clay of bed No. 6, has a thick-slaty structure, a grayish-black colour, and a shining resinous streak. It is sectile, but does not yield to the nail. Pieces of lignite occur imbedded in it, and it is traversed by fibrous ramifications of carbonaceous matter.

Specimens 115, 116, 117, 118, 119, are of substances altered by contact with beds of burning coal.

Mr. Drummond brought specimens from the spot which Mr. Small alludes to and remarks, that the lignite occurs in beds from six inches to two feet thick, separated by clay and sandstones. His specimens of the lignite

1051, 1052, 1053 are precisely similar to the slaty and conchoidal varieties, which occur at the mouth

1055 of the Bear Lake River; and there is an equal resemblance betwixt the sandstones from the two places.

1053 The slaty beds of lignite, at Edmonton, pass into a thin, slaty, friable sandstone, much impregnated by carbonaceous matter, and containing pieces of fibrous

1056, 1062 lignite. In the neighbourhood of the lignite there are some beds of rather indurated, but highly bituminous shale, and the clayey banks contain clay-iron stones, in form of septaria. Mr. Drummond likewise found beds of a beautiful bituminous coal, which Professor Buckland, from its peculiar fracture, considers to be tertiary pitch-coal. The banks of the Saskatchewan, near the same place, exhibit beds of a very compact stone, having a brown colour, and inclosing many fragments of bituminous limestone and some organic remains; like-

1058, 1059, 1060 wise beds of a somewhat similar stone, but full of drusy cavities, and more resembling a recent calcareous tufa. I could not learn how far these beds were connected with the lignite deposit.

Captain Franklin* saw beds of lignite and tertiary pitch-coal at Garry's Island, off the mouth of the Mackenzie, and there is an extensive deposit of it near the Babbage River, on the coast of the Arctic Sea, opposite to the termination of the Richardson chain of the Rocky Mountains.

MACKENZIE RIVER FROM SLAVE LAKE TO THE BASE OF THE ROCKY MOUNTAINS.

Having now described the strata in Bear Lake River, together with the exposed beds of the lignite at its mouth, as far as opportunities of observation enable me, and also added a slight account of similar formation which occupy a like situation at the foot of the Rocky Mountain range, were I to adapt the order of my notices strictly to the route of the expedition, I should next describe the banks of the Mackenzie from the junction of the Bear Lake River downwards to the Arctic Sea. It seems, however, more advisable to commence at the

* See Page 50 of the Narrative.

origin of the Mackenzie, in Great Slave Lake, and give as connected a view as I can of the principal geological features of that great river.

The west end of Slave Lake is bounded by horizontal strata of a limetone, whose characters shall be afterwards given in detail; and I have merely to remark, at present, that it forms flat shores, which are skirted by shoals of boulders of limestone, and of primitive rocks. Much drift timber is accumulated in the small bays at this end of the lake, which, in process of time, is converted into a substance like peat. A chain of islands extends obliquely across the lake at the origin of the river, or where the current is first felt; and the depth of the water there is less than six feet. Below this, there is a dilatation termed the *first little lake*, and the river afterwards contracts to less than a mile in breadth; forming in one place, when the water is low, a strong rapid. A second dilatation, about twenty-five miles below the first, is termed the *second little lake.* The shores throughout this distance are generally flat and covered with boulders of limestone, compact felspar, granite, gneiss, and sienite, and there are many of these stones imbedded in a tenacious clay, which forms the beach. A ridge, having an even outline, and apparently of small elevation, commences behind Stony Point, in Slave Lake, some distance inland, and, running nearly parallel to the river, disappears about Fishing River, a stream which joins the Mackenzie, below the Second Little Lake. The Horn Mountains, a ridge of hills, of considerably greater elevation, and having a more varied outline than that on the south shore, are first visible on the north side of the Second Little Lake, and continue in sight nearly as far as the junction of the "River of the Mountains," or "Forks, of the Mackenzie," as the traders term the union of the two rivers. The only rocks seen
120 *in situ* between Slave Lake and the Forks, were a bitu-
121 minous shale of a brownish-black colour, in thin slates, and a slate-clay of a pure yellowish-gray colour, which, as well as the bituminous shale, forms steep banks.

ROCKY MOUNTAINS.

About twenty-five or thirty miles below the forks, the first view is obtained of the Rocky Mountains, which there appear to consist of short-conical peaks, scarcely rising two thousand feet above the river. Some distance lower down, the river, changing its course from W.N.W. to N.N.E., turns sharply round the mountains, which are there disposed in ridges, having bases from one to two miles wide, and a direction of S.S.W. or S.W. by S. being nearly at right angles to the general course of the great range to which they belong. The eastern sides of the ridges present a succession of wall-sided precipices, having beneath them shelving acclivities formed by debris, and exhibiting on their faces regular lines of stratification. The western sides of the ridges are of more easy ascent. The vallies which separate these ridges and open successively to the river, are narrow, with pretty level bottoms, but very steep sides well clothed with trees. In the first ridge, the strata seemed to dip to the northward at an angle of 35°. In some of the others they were horizontal, or had a southerly dip. The third ridge presents, when viewed from the westward, a magnificent precipice, seemingly about one thousand two hundred feet high, and which extends for at least fifteen miles.* After passing this ridge, the river inclines to the eastward, and the forms of the hills are less distinctly seen.

As I could not visit the Rocky Mountains, I know nothing of their structure except from report. An interpreter in the Hudson's Bay Company's service, who had travelled over them, informed me that there are fourteen or fifteen ridges, of which the three easternmost are the most rugged, those that succeed being broader and more rounded. This man gave me a

122 specimen of a pearl-gray semi-opal, resembling obsidian, brought from the third or fourth ridge. The natives, by means of fire, cause this stone to break off in thin, flat, conchoidal fragments, with which they form arrow-heads and knives. The thin pieces are nearly transparent on the edges.

He also gave me a specimen of plumbago, from the same

123 quarter, and some specular iron.

Mr. Macpherson, of the Hudson's Bay Company, in a letter respecting the Rocky Mountains, near *Fort au Liard*, on the River of the Mountains, or south branch of the Mackenzie, informs me, that "these mountains may be traced into somewhat uniform ranges, extending north-westerly and south-easterly, nearly parallel with the River of the Mountains, and are in appearance confusedly scattered and broken, rising here and there into high peaks." This gentleman had the kindness to send me specimens of a cherty rock, some of which,

124, 125 he states, were from the third range westward from the river, and others from a spur which projects in a southern direction from the fourth range, and rises about six hundred feet above the adjacent valley. These specimens cannot be distinguished from those of Limestone Point, on the north shore of Great Bear Lake, (noticed in page 267.)

Mounts Fitton and Conybeare, two remarkable peaks which terminate the Eastern range of the Rocky Mountains on the shores of the Arctic sea, were found by Captain Franklin to consist of transition rocks, of which an account is given in the subjoined note.*

* List of specimens, collected by Captain Franklin, on the sea-coast, to the westward of the Mackenzie.

From Mount Fitton in the Richardson Chain.

344 Grauwacke-slate in columnar concretions, detached from the rocky strata by an Esquimaux.
348 Grauwacke-slate, resembling the preceding, from the same place. Used by the Esquimaux as a whetstone.
345, 346 Globular balls of dark, blackish-gray, splintery limestone, and of flinty-slate, traversed by minute veins of calc-spar. Picked up at the base of the mountain.
347 Worn pebbles of quartz, lydian stone, splintery limestone, and grauwacke, from the same spot.
349 Fine-grained, mountain-green clay-slate, approaching to potstone; quarried by the Esquimaux in the Cupola Mountain of the same chain, and used to form utensils.
350 Rock-crystal from the same chain of mountains.

From the beach between Point Sabine and Point King.

351 Brown-coal, woody structure scarcely perceptible. There are beds of this coal in the earthy cliffs where the party was encamped on the 13th and 14th July near Point King.
352 Clay-iron stone, forming boulders in the channels of the rills, which cut the earthy banks containing coal.
353, 354 Pitch-coal, having a fibrous structure and a very beautiful fracture, presenting a congeries of circles. (This coal was recognised by Professor Buckland to be a tertiary pitch-coal, and is precisely similar to specimens brought from the upper branches of the Saskatchewan, by Mr. Drummond: see page 284.) The specimen was picked up from the gravelly beach at the mouth of the Babbage River.
355 Greenish-gray limestone, with a somewhat earthy granular aspect; containing shells which Mr. Sowbery considers to be very like the *cyclas medius* of the Sussex weald-clay. Picked up at the same place with the preceding specimen.

Captain Franklin remarks, that "the Babbage flows between the mountains of the Richardson Chain, and that there were no solid strata nor any large boulders near its mouth. The gravel consisted of pebbles of red and white sandstone, slaty limestone, greenstone, and porphyry, much worn by attrition."

From Mount Conybeare, in the Buckland Chain.

356 Greenish-gray grauwacke slate, (resembling No. 348,) with specks of effervescent carbonate of lime. The surfaces of the slates exhibit interspersed scales of mica. The specimens were broken from the summit of Mount Conybeare, at the western extreme of the Buckland Chain: latitude 69° 27′, longitude 139° 53′ west.
358 Fine-grained grauwacke-slate in columnar concretions, from the same place with specimen 356.
357 Grauwacke-slate, in thick slaty columnar concretions, besprinkled with

Sir Alexander Mackenzie, towards the conclusion of the interesting narrative of his voyages, says, of the Rocky Mountain range, "The last line of division is, the immense ridge, or succession of ridges of the stony mountains, whose northern extremity dips in the Arctic Sea in latitude 70° north, and longitude 135° west, running nearly south-east, and begins to be parallel to the coast of the Pacific ocean from Cook's inlet, and so onwards to the Columbia. From thence it appears to quit the coast, but still continuing with less elevation to divide the waters of the Atlantic from those of the Pacific. In these

scales of mica. Taken from a bed about the middle of Mount Conybeare. The resemblance of this stone to that of Mount Fitton (No. 344) is very remarkable.

360 Similar rock to 358, with an adhering portion of a vein of crystallized quartz, and on one side a bit of bluish-gray slate. From the middle of Mount Conybeare.

359 Columnar concretion of a slaty rock, like 356, but more quartzose, breaking into rhomboidal fragments. From the middle of Mount Conybeare.

361, 362 Grauwacke-slate, with a thin adhering vein of carbonate of lime and numerous particles of disseminated mica. From the middle of Mount Conybeare.

363 Bluish-gray grauwacke-slate, resembling Nos. 348 and 344. From the Upper Terrace, at the base of Mount Conybeare.

364 Dark-bluish gray and very fine-grained grauwacke-slate, with a glimmering lustre, traversed by a vein of quartz. From the same place.

365 A thick-slaty angular concretion of a very quartzose grauwacke-slate, (similar to Nos. 348 and 358,) decomposed on the surface and breaking into rhomboidal fragments. From the middle Terrace at the base of Mount Conybeare.

366 A somewhat rhomboidal portion of flinty-slate, apparently part of a bed. From the Lower Terrace of Mount Conybeare, which is composed of this rock. The terrace is ten miles distant from the sea-coast, and the intervening ground is swampy.

The whole series of specimens from Mount Conybeare, (Nos. 356 to 366,) appear to belong to transition rocks; and the continuity of the formation with that of Mount Fitton is rendered probable, both by the resemblance of the specimens and the geographical situation of the mountains.

Captain Franklin saw no rocks, *in situ*, on the coast to the westward of the Richardson Chain; but he gathered boulders of the following rocks from the bed of the Net-setting Rivulet, which flows from the British Chain of the Rocky Mountains, and falls into the Arctic Sea, between Sir P. Malcolm River and Backhouse River.

367 Greenstone; 368, yellowish-gray sandstone; 369, dark-coloured splintery-limestone; 370, 371, 372, dolomite; 373, quartzose sandstone, like the old red sandstone; 374, grauwacke-slate; 375, quartz and iron pyrites.

Boulders of the under-mentioned rocks were gathered on Flaxman Island.

378 Fine-grained, greenish clay-slate, obviously of primitive rock, abundant in the neighbourhood, and supposed to have been brought down by the rivulets which flow from the Romanzoff Chain. 379, quartz.

376 and 377 were from Foggy Island, and are rolled specimens of flinty-slate; one of them containing corallines.

snow-clad mountains rises the Mississippi, if we admit the Missouri to be its source, which flows into the Gulph of Mexico; the river Nelson which is lost in Hudson's Bay; Mackenzie's river that discharges itself into the North Sea, and the Columbia emptying itself into the Pacific Ocean. The breadth of the mountains from Cook's inlet to the Columbia is from four to eight degrees easterly." I may add, that the great rivers mentioned by Mackenzie not only take their origin from the same range of mountains, but almost from the same hill; the head waters of the Columbia and Mackenzie being only about two hundred yards apart in latitude 54½°. Mr. Drummond, who crossed the mountains at that place, informs me, that the Eastern side of the range consists of conglomerate and sandstone, to which succeed limestone hills exceedingly barren, and afterwards clay-slate and granite.

James, the intelligent naturalist, who accompanied Major Long on his first expedition, says of the Rocky Mountains to the southward of the Missouri, "They rise abruptly out of the plains which lie extended at their base on the east side, towering into peaks of great height, which renders them visible at the distance of more than one hundred miles from their base. They consist of ridges, knobs, and peaks, variously disposed, among which are interspersed many broad and fertile valleys. James's peak, one of the more elevated, was ascertained by trigonometrical measurement to rise 8500 feet above the common level. The rocky formations are uniformly of a primitive character, but a deep crust of secondary rocks appears to recline on the east side of the mountains, extending upwards from their base many hundred feet." In another place, he says, "The woodless plain is terminated by a range of naked and almost perpendicular rocks, visible at the distance of several miles, and resembling a vast wall parallel to the base of the mountain. These rocks are sandstone, and rise abruptly to an elevation of one hundred and fifty or two hundred feet." The sandstone walls seem to present an appearance not very dissimilar to some of the cliffs seen from the Mackenzie.

Having thus mentioned as briefly as I could the extent of the information I was able to collect, respecting the Rocky Mountain range, I may remark, that a formation of primitive rocks, but little elevated above the general level of the country, appears to run from near the west end of Lake Superior, gradually and slightly converging towards the Rocky Mountains, until it attains the east side of Great Bear Lake. In lat. 50°, the two ranges are nearly seven hundred miles apart, and there, and as far as lat. 60°, the space between them is principally occupied by horizontal strata of limestone. There is also much limestone in the narrower interval north of 60°, but the strata are more inclined, and form abrupt hills and ridges, particularly about lat. 66°, where the primitive rocks on the east of Bear Lake are within two hundred miles of the Rocky Mountains. Sir Alexander Mackenzie has noticed that a chain of great lakes skirts this eastern range of primitive rocks, where they are approached by the flat limestone strata which lie to west of them. Thus the primitive rocks bound Great Slave Lake to the eastward of Slave River, and the flat limestone strata occupy the country westward of that lake, as has been already mentioned.

After this digression, which seemed necessary for the purpose of giving a general idea of the structure of the country, I return to the description of the banks of the Mackenzie.

MACKENZIE RIVER FROM THE FIRST SIGHT OF THE ROCKY MOUNTAINS TO BEAR LAKE RIVER.

At the sharp turn of the river round the Rocky Mountains, its east bank swells gently into a hill several feet high. Below this the banks are broken into conical masses by ravines, and present a finely variegated outline. A pretty high ridge, looking like a continuation of the Horn Mountain, is visible on the east side some distance inland. Opposite to the Big Island there is a green hill three or four hundred feet high, which, as we descended the river, showed itself to be part of a range that had a direction apparently to the N.N.W., and towards its northern end became more rugged and craggy, exhibiting cliffs and rude embrasures, at the same time increasing in height to eight hundred or one thousand feet. The boulders on the beach change their character considerably about this place. Farther up, the yellowish-white limestone which occurs in Slave Lake formed a great portion of them; but here a greenish-gray, and rather dark-coloured, compact limestone, with a flat conchoidal fracture, replaces it. Variegated-sandstone, and some purplish, felspathose-sandstone, or compact felspar, also occur pretty frequently, together with slaty limestone, bituminous-shale, lydian-stone, pitchstone-porphyry, and various sienities, granites, and greenstones, almost all porphyritic.

The Rock by the river's side presents the first solid strata that occur on the immediate banks of the river after passing the Forks. It is a round bluff hill about five hundred feet high, with a short obtuse-conical summit. A precipice three hundred feet high, washed by the river, is composed of strata of limestone, dipping N.W. by W. at an angle of 70°; but the strata in other parts of the hill have in appearance the saddle-formed arrangement. The limestone is of a blackish-
127 gray colour, slightly crystalline structure, and much resembles the stone of the principal beds in the hills at the rapid and mouth of Bear Lake River. Its beds are from one to two feet thick, and much intersected by small veins of calc-spar. There are also some larger veins a foot and a half thick, which traverse the strata obliquely, having their sides lined with calc-spar, and their centres filled with transparent gypsum. I observed a small imbedded pebble of white
128 sandstone in the gypsum. Some of the beds of limestone
127 consist of angular distinct concretions. A small island lying off this rock, having its strata dipping south at an
131 angle of 20°, presents a bed a foot thick, entirely com-
132 posed of these angular concretions, covered by a thin-slaty limestone, and reposing on thicker beds, all of which are dark-coloured. No organic remains were observed.

A few miles below the "Rock by the river side," a very rugged ridge appears on the eastern bank. It has sharp craggy summits, and is about five or six hundred feet high. For nearly sixty miles below this place the river continues about eight hundred yards wide, bounded by banks chiefly of clay; but in some places of a clayey shale having a bluish colour. The banks are in many places one hundred and fifty feet high, with a beach beneath covered with boulders. A little above the site of the Old Fort Norman the river dilates, and is full of islands; and a short way inland, on the east side, stands Clark's Hill, which is visible from Fort Franklin, and is supposed to be near 1500 feet high. It is shaped somewhat like the amphibolic-granite mountain of Criffel in Galloway, and in its immediate neighbourhood there are some less lofty, but very rugged and precipitous hills, resembling in outline the ridges of limestone on Bear Lake River. From this place to the commencement of the lignite formation, already described, the banks of the Mackenzie are high and clayey.

MACKENZIE RIVER FROM BEAR LAKE RIVER TO THE NARROWS.

Below Bear Lake River the general course of the Mackenzie for eighty miles is about N.W. by W., when a remarkable rapid is produced by ledges of stone which cross its channel. The width of the river varies in this distance from one to three miles, but the water-course is narrowed by numerous islands, and the current continues strong. The Rocky Mountains are visible, running in a direction from S.E. to N.W. Judging merely by the eye, we did not estimate their altitude above four thousand feet, and I may remark, that the snow disappears from their summits early in the summer. A back view of the hill at the mouth of Bear Lake River is also obtained for upwards of twenty miles, but the ridge of which it forms a part curves inland, probably uniting, as was formerly remarked, with the one which crosses Bear Lake River near the middle of its course. The banks of the Mackenzie are in general from one hundred and twenty to one hundred and fifty feet high in this part, and there are occasional sections of them, but we had little leisure to examine their structure. In the voyage of 1826 we drifted down the stream night and day, landing only when necessary to cook our provisions; and in the following geological notices, as far as the passage of the river named the *Narrows*, I have done little more than describe the specimens collected by Captain Franklin, when he ascended the river by the tow-line in 1825. The few notes that the rapidity of our voyage permitted me to make, as to the direction of the strata, &c., were inserted in the book that was purloined by the Esquimaux at the mouth of the river.

About fifty miles below Bear Lake River there is an almost precipitous cliff of bituminous-shale, one hundred and twenty feet high, strongly resembling the cliffs which occur near the bases of the hill of Scented-Grass and Great Bear Mountain in Bear Lake already described*, and at the mouth of the Clear Water River in lat. 56½°. In the two former localities the shale is in the neighbourhood of horizontal strata of limestone; and in the latter it actually reposes on the limestone, which extends in horizontal strata as far as Great Slave Lake, is connected with many salt springs, and possesses many of the

* Page 268.

characters ascribed to the zechstein formation. Captain Franklin observed the beach under the shale cliffs of 133 the Mackenzie to be strewed not only with fragments of the shale, but also with much lignite, similar to that which occurs at the mouth of the Bear Lake River. Twelve or fourteen miles below these cliffs there is a reach seventeen or eighteen miles long, bounded by walls of sandstone in horizontal beds. Specimens obtained by Captain 134 Franklin at the upper end of the reach consist of fine-135 grained quartzose sandstone* of a gray colour, and having a clayey basis, resembling those which occur in the middle of Bear Lake River. At the commencement of the "Great Rapid of the Mackenzie" there is a hill on each side of the river, named by Captain Franklin the eastern† and Western mountains of the Rapid. The Rocky Mountains appear at no great distance from this place, running about N.W. by W., until lost to the sight; and as the Mackenzie for forty or fifty miles below, winds away to the northward, and, in some, reaches a little to the eastward, they are not again visible, until the river has made a bend to the westward, and emerges from the defile termed "the Narrows."

The "Eastern mountain of the rapid" seems to have a similar structure, with the "Hill by the River's side," the hill at the mouth of Bear Lake River, and the other limestone ridges which traverse this part of the country. From some highly inclined beds near its base I broke off specimens of a lime-136 stone, having an imperfectly crystalline structure, and a brown colour, which deepens into dull black on the surfaces of its natural seams. A piece of dark-gray, compact limestone, having the peculiar structure to which the 137 name of "*cone in cone*" has been given, was found on the 138 beach; also several pieces of chert, and some fragments

* 134. These specimens have a wood-brown colour internally, and appear to be composed of minute grains of quartz, variously coloured, white, yellowish-brown and black, cemented together by an earthy basis. It is a hard and apparently durable stone, occurring in layers an inch thick, and having its seam-surfaces of a grayish-black colour, with little lustre, as if from a thin coating of bituminous clay.

135, are specimens of a more compact, harder, and finer-grained quartzose sandstone, with less cement, and of a deeper bluish-gray colour.

† Mackenzie attempted to ascend this hill, but was compelled to desist by clouds of musquitoes, (July 6th, 1789. *Voyage to the Arctic Sea*, p. 40.)

136 This limestone effervesces strongly with acids, breaks into irregular fragments, but with an imperfect slaty structure, and has a brown colour, with considerable lustre in the cross fracture.

139 of a trap-rock, consisting of pieces of greenstone, more
141 or less iron-shot, cemented by calc-spar.

Immediately below the rapid there are horizontal layers of sandstone which form cliffs, and also the bed of the river. Captain Franklin obtained specimens of this stone, which do not differ from the sandstones above the rapid. And
142 amongst the debris of the cliff he found other specimens
143 of the "*cone in cone*," such as it occurs in the clayey
140 beds of the coal measures, and also some pieces of crystallized pyrites.

About forty miles below the rapid, the river flows through a narrow defile formed by the approach of two lofty banks of limestone in highly-inclined strata, above which there is
144 a dilatation of the river, bounded by the walls of sand-
144a stone, which have weathered, in many places, into pil-
145 lars, castellated forms, caves, &c. The sandstone strata
146 are horizontal, have slate-clay partings, and seams of a
147 poor clay-iron stone, but do not differ in general appearance from the sandstone beds at the rapid, except that a
144b marly stone containing corallines, and having the general colour and aspect of the sandstone beds, is associated with them at this place.

The very remarkable defile, below these sandstone beds, is designated "the *second rapid*" by Sir Alexander Mackenzie, and "the *ramparts*" by the traders, a name adopted by Captain Franklin. Mackenzie states it to be three hundred yards wide, three miles long, and to have fifty fathoms depth of water. If he is correct in his soundings, its bed is probably two hundred and fifty feet below the level of the sea. The walls of the defile rise from eighty to one hundred and fifty feet above the river, and the strata are inclined to the W.N.W., at an angle of seventy or eighty degrees. It is worthy of remark,

The specimens collected by Captain Franklin were as follows:—

144a Sandstone of an ash-gray colour, composed of rounded grains of semi-transparent quartz of various sizes, imbedded in a considerable proportion of a powdery basis which effervesces with acids. This bed weathers readily.

145 Thick-slaty sandstone passing into slate-clay, having a very-fine grained earthy fracture, and a light bluish-gray colour. It is very similar to some of the softer sandstones that occur in the coal field at Edinburgh, particularly in the Calton Hill.

146 Sectile ash-gray slate-clay which forms the partings of the beds.

144b Bluish-gray marl, impregnated with quartz, forming a moderately hard stone, and containing corallines (*amplexus.*)

that the course of the river through this chasm is E.N.E., and that just above the eastern mountain of the rapid it runs about W.S.W. through the sandstone strata, as if it had found natural rents by which to make its escape through the ridge of hills which cross its course here. Similar elbows occur in various parts of the River, and they may be almost always traced to some peculiarity in the disposition of the hills which traverse the country.

Captain Franklin gathered many specimens of the limestone strata of the Ramparts, which are specifind in a note.* Some of the beds at the upper part of the Ramparts consist of a granular foliated limestone, which was not noticed elsewhere on the banks of the river, but the greater part are of limestone, strongly resembling that which has been already described, as forming the ridges in this quarter. Most of the beds are impregnated wholly, or in

148, 149

* *Upper part of the ramparts.*

148 A fine-granular, foliated limestone, of a white colour, having large patches stained yellowish-brown, apparently by bitumen.
149 A yellowish-gray slightly granular limestone, with disseminated calc-spar.
150 Compact, white limestone, which, when examined with a lens, appears to be entirely composed of madrepores.
151 Specimens of limestone, having a crystalline texture, a brownish colour and slaty structure.
152 The seams are dark, as if from the carbonaceous matter—portions of this bed have the appearance of old mortar; but contain obscure madrepores.

From the middle of the ramparts.

153 Fine-granular limestone, having a pale, wood-brown colour, and a splintery fracture. It resembles the limestone of the hill at the mouth of Bear Lake River.
154 Pale yellowish-brown limestone, with a dull fracture, but interspersed with small, shining, sparry plates, and traversed by concretions of calc-spar, that appear to have originated in corallines.
155 Yellowish-gray limestone, passing into a soft marl slate.
156 Some beds contain a shell, which Mr. Sowerby refers, though with doubt, to the species named terebratula sphœroidalis, a fossil of the corbrash. The substance of the shells is preserved.

Some of the specimens contain *producti*, and fragments of the coral named *amplexus*.

Lower end of the ramparts.

157 Fine-grained limestone, of a dark-brown colour, containing some small, round, smooth balls of dark limestone—occurs in horizontal strata.
158 Brownish-black flinty-slate, which forms a layer an inch thick, and covers the horizontal beds of limestone last mentioned. (157.)

patches, with bitumen. Some of these specimens contain corallines and terebratulæ; and at the lower end of the defile there are horizontal strata of limestone, covered by a thin layer of flinty slate.

Below the *ramparts* the river expands to the width of two miles, and for a reach or two its banks are less elevated. In lat. 66¾°, about thirty miles from the ramparts, there are cliffs which Captain Franklin in his notes, remarks, "run on an E. by S. course for four miles, are almost perpendicular, about one hundred and sixty feet high, and present the same castellated appearances that are exhibited by the sandstone above the defile of the "ramparts." The cliffs* are, in fact, composed of sandstones similar, in general appearance, to those which occur higher up the river; but some of the beds contain the quartz in coarser grains, with little or no cement. The beds are horizontal, and repose on horizontal limestone,† from which Captain Franklin broke many specimens in 1825. We landed at this place in 1826 to see the junction of the two rocks, but the limestone was concealed by the high waters of the river.

159, 160, 161, 162, 163, 164, 165, 166, 167, 168, 169, 170,

* *Specimens from the cliffs in lat.* 66¾°.

159 Very fine-grained sandstone, with much clayey basis—portions of the bed iron-shot.

160 Sandstone fine-grained, and appearing, when examined with a lens, to be composed of minute grains of whitish translucent quartz, black Lydian stone, and ochre-coloured grains, probably of disintegrated felspar.

161 Rounded grains of nearly transparent quartz united without cement—this stone is friable.

162 Sandstone composed of grains like the preceding, united by a basis, and forming a firmer stone.

163 Hard, thin, slaty, bluish-gray sandstone, much iron-shot.

164 Fine-grained bluish-gray sandstone, not to be distinguished in hand-specimens from some of the sandstones which occur at the rapid in Bear Lake River.

† *Horizontal limestone beds lying under the sandstone.*

166 Fine-grained limestone, with an earthy fracture, coloured brown and grayish-white in patches.

167, 168 Similar stone to preceding, containing many shells. Some beds contain only broken shells.

169 Bed of imperfectly crystalline limestone, of a brownish-gray colour, traversed by veins of calc-spar.

170 Fragments containing madrepores and chain coral—occur amongst the debris of the limestone cliffs.

Captain Franklin's specimens are full of shells, many of which are identical with those of the flat limestone strata of the Athabasca River. One bed appears to be almost en-
171 tirely composed of a fine large species of terrebratula, not yet described, but of which Mr. Sowerby has a specimen from the carboniferous limestone of Neho, in Norway. Some of the beds contain the shells in fragments; in others, the shells are very entire.

About forty miles below these sandstone walls the banks of the river are composed of marl-slate, which weathers so readily, that it forms shelving acclivities. In one reach the
172 soft strata are cut by ravines into very regular forms, resembling piles of cannon shot in an arsenal, whence it was named *Shot-reach.*

The river makes a short turn to the north below Shot Reach, and a more considerable one to the westward, in passing the present site of Fort Good Hope. The banks in that neighbourhood are mostly of clay, but beds of sandstone occasionally show themselves. The Indians travel from Fort Good Hope nearly due north, reach the summit of a ridge of land on the first night, and from thence following the course of a small stream they are conducted to the river *Inconnu*, and on the evening of the 4th day they reach the shores of Esquimaux Lake. Its water is brackish, the tide flowing into it. The neck of land which the Indians cross from Fort Good Hope is termed "isthmus" on Arrowsmith's map, from Mackenzie's information; and its breadth, from the known rate at which the Indians are accustomed to travel, cannot exceed sixty miles. The ridge is named the Carrebœuf, or Rein-deer Hills, and runs to lat. 69°, forming a peninsula between the eastern channel of the Mackenzie and Esquimaux Lake.

A small stream flows into the Mackenzie some way below Fort Good Hope, on the banks of which, according to Sir Alexander Mackenzie, the Indians and Esquimaux collect flints. He describes these banks as composed of "a high, steep, and soft rock, variegated with red, green, and yellow hues; and that, from the continual dripping of the water, parts of it frequently fall, and break into small, stony flakes, like slate, but not so hard. Amongst these are found pieces of petroleum, which bears a resemblance to yellow wax, but is more pliable." The flint he speaks of is most probably flinty-slate; but I do not know what the yellow petroleum is, unless it be the variety of alum, named rock-butter, which was observed in other situations, forming thin layers in bituminous shale.

About twenty miles below Fort Good Hope there are some sandstone cliffs,* which Captain Franklin examined 173, 174 in 1825. The sandstones are similar to those occurring higher up the river, but some of the beds contain small pieces of bituminous shale; and they are interstratified with thin layers of flinty-slate, and of flinty-slate passing into bituminous shale. The flinty-slate contains 175, 176 iron pyrites, and its layers are covered with a sulphureous efflorescence. Some of the beds pass into a slate-clay, which contains vegetable impressions, and some veins of clay-iron stone also appear in the cliff.

Sixty miles below Fort Good Hope the river turns to the northward, and makes a sharp elbow betwixt walls of sandstone eighty or ninety feet high, which continue for fifteen or twenty miles. Captain Franklin named this pas- 178 sage of the river "The Narrows."† The sandstones 179 of the *Narrows* lie in horizontal beds, and have generally a dark gray colour. They are parted by 180 thin slaty beds of sandstone, containing small pieces 181, 182 apparently of bituminous coal, and some casts of 183 vegetables. Most of the beds contain scales of mica, and some of them have nodules of indurated ironshot clay which exhibit obscure impressions of shells. A bed of imperfectly crystalline limestone was seen by Captain Franklin underlying the sandstones.

* *Sandstone cliffs twenty miles below Fort Good Hope.*

173 Friable sandstone, composed of grayish-white quartz, in smooth, rounded grains, cemented by a brownish basis. Some carbonaceous matter is interspersed through the stone, and it contains small fragments of bituminous shale.

174 Calcareous sandstone passing into slate-clay—bluish-gray colour.

175 Black, flinty-slate, with a flat conchoidal cross fracture. Some of the pieces appear to be rhomboidal distinct concretions.

176 Dull, flinty-slate, with an even fracture.

178 Thin-slaty blackish-gray sandstone, much indurated, containing scales of mica.

179, 180 Bluish-gray sandstone, containing many minute specks of carbonaceous matter; also, in patches, grains of chert, and flinty-slate, and imbedded pieces of iron-shot clay, which has obscure casts of shells. Scales of mica are interspersed through this stone.

181, 182 Sandstone containing specks of bituminous? coal, and casts of some vegetable? substance.

183 Gray limestone, much impregnated with quartz, and having an imperfect crystalline structure.

† Mackenzie notices the precipices of "gray stone," which bound the river here, p. 71.

MACKENZIE RIVER BELOW "THE NARROWS."

THE Mackenzie, on emerging from the Narrows, separates into many branches, which flow to the sea through alluvial or diluvial deltas and islands. The Rocky Mountains are seen on the western bank of the river, forming the boundary of those low lands; and the lower, but decided ridge, of the Rein-deer Hills holds nearly a parallel course on the east bank. The estuary lying between these two ranges, opens to the N.W. by N. into the Arctic Sea. I have already mentioned the specimens of rocks obtained at the few points of the Rocky Mountains that were visited,* and therefore shall now speak only of the Rein-deer Hills. We did not approach them until we had passed for thirty miles down a branch of the river which winds through alluvial lands. At this place there are several conical hills about two hundred feet high, which appeared to consist of limestone. Specimens taken from some slightly-inclined beds near their bases, consisted of a fine-grained, dark, bluish-gray limestone. After passing these limestone rocks, the Rein-deer Hills were pretty uniform in appearance, having a steep acclivity with rounded summits. Their height, on the borders of the river, is about four hundred feet, but a mile or two inland they attain an elevation of perhaps two hundred feet more. Their sides are deeply covered with sand and clay, arising most probably from the disintegration of the subjacent rocks. A section made by a torrent, showed the summit of one of the hills to be formed of gray slate-clay, its middle of [184, 185] friable gray sandstone much iron-shot, and its base of dark bluish-gray slaty clay. The sandstone predominates in some parts of the range, forming small cliffs, underneath which there are steep acclivities of sand. It contains nearly an equal quantity of black flinty slate, or lydian stone, and white quartz in its composition, and greatly resembles the friable sandstones of the lignite formation at the mouth of Bear Lake River. In some parts the soil has a red colour [186] from the disintegration of a reddish-brown slate-clay. The summits of the hills that were visited were thinly [187] coated with loose gravel, composed of smooth pebbles of lydianstone, intermixed with some pieces of green felspar, white quartz, limestone, and chert. In some places almost

* See page 283.

all the pebbles were as large as a goose-egg; in others none of them exceeded the size of a hazel nut. The Rein-deer Mountains terminate in lat. 69°, having previously diminished in altitude to two hundred feet, and the eastern branch of the river turns round their northern extremity. White spruce trees grow at the base of these hills as far as lat. 68½° ; north of which they become very stunted and straggling, and very soon disappear, none reaching to lat. 69°.

Sir Alexander Mackenzie, who, on his return from the sea, walked over these hills, says, "Though the country is so elevated, it is one continued morass, except on the summits of some barren hills. As I carried my hanger in my hand, I frequently examined if any part of the ground was in a state of thaw, but could never force the blade into it beyond the depth of six or eight inches. The face of the high land towards the river is, in some places, rocky, and in others a mixture of sand and stone, veined with a kind of red earth, with which the natives bedaub themselves." It was on the 14th of July that he made these observations. On the 5th of the same month, in a milder year, we found that the thaw had penetrated nearly a foot into the beds of clay at the base of the hills.

ALLUVIAL ISLANDS AT THE MOUTH OF THE MACKENZIE.

The space between the Rocky Mountains and Rein-deer Hills, ninety miles in length from lat. 67° 40′ to 69° 10′, and from fifteen to forty miles in width, is occupied by flat alluvial islands, which separate the various branches of the river. Most of these islands are partially or entirely flooded in the spring, and have their centres depressed and marshy, or occupied by a lake; whilst their borders are higher and well clothed by white spruce trees. The spring floods find their way, through openings in these higher banks, into the hollow centres of the islands, carrying with them a vast quantity of drift timber, which, being left there, becomes water-soaked, and, finally, firmly impacted in the mud. The young willows, which spring up rapidly, contribute much towards raising the borders of the stream, by intercepting the drift sand which the wind sweeps from the margin of the shallow ponds as they dry up in summer. The banks, being firmly frozen in spring, are enabled to resist the weight of the temporary floods which occur in that season, and before they are thawed the river has resumed its low summer level. The trees which grow on the islands terminate suddenly, in lat. 68° 40′.

I have already mentioned, that a large sheet of brackish water, named Esquimaux Lake, lies to the eastward of the Rein-deer Mountains, running to the southward, and approaching within sixty miles of the bend of Mackenzie's River at Fort Good Hope. This lake has a large outlet into Liverpool Bay, to the westward of Cape Bathurst, and there are many smaller openings betwixt that bay and Point Encounter, near the north end of the Rein-deer Hills, which are also supposed to form communications betwixt the lake and the sea. The whole coast-line from Cape Bathurst to the mouth of the Mackenzie, and the islands skirting it, as far as Garry and Sacred Islands, present a great similarity in outline and structure. They consist of extensive sandy flats, from which there arise, abruptly, hills of an obtuse conical form, from one to two hundred feet above the general level. Sandy shoals skirt the coast, and numerous inlets and basins of water divide the flat lands, and frequently produce escarpments of the hills, which show them to be composed of strata of sand of various colours, sometimes inclosing very large logs of drift timber. There is a coating of black vegetable earth, from six inches to a foot in thickness, covering these sandy hummocks, and some of the escarped sides appeared black, which was probably caused by soil washed from the summit.

It is possible that the whole of these eminences may, at some distant period, have been formed by the drifting of moveable sands. At present the highest floods reach only to their bases, their height being marked by a thick layer of drift timber. When the timber has been thrown up beyond the reach of ordinary floods, it is covered with sand, and, in process of time, with vegetable mould. The *Elymus mollis*, and some similar grasses with long fibrous roots, serve to prevent the sand-hills from drifting away again. Some of the islands, however, consist of mud or clay. Captain Franklin describes Garry's Island as presenting cliffs, two hundred feet high, of black mud, in which there were inclined beds of lig-
188 nite. Specimens of this lignite have the same appearance with the fibrous wood-coal occurring in the formation at the mouth of Bear Lake River, and, like it,
189 contain resin. Imbedded in the same bank, there were
190 large masses of a dark-brown calc-tuff, full of cavities containing some greenish earthy substance. Some boulders of lydian stone strew the beach. The cliffs of Ni-

cholson's Island also consisted of sand and mud, which, at the time of our visit, (July 16th,) had thawed to the depth of three feet. This island rises four hundred feet above the level of the sea, and is covered with a thin sward of grasses and bents.

SEA-COAST.—BITUMINOUS ALUM SHALE.

The main land to the east of Nicholson's Island, as far as Cape Bathurst, presents gently swelling hills, which attain the height of two hundred feet at the distance of two miles from the beach, and the ground is covered with a sward of moss and grasses. At Point Sir Peregrine Maitland there are cliffs forty-feet high of sand and slaty clay, and the ravines are lined with fragments of whitish compact limestone, exactly resembling that which occurs in Lakes Huron and Winipeg, and which was afterwards seen forming the promontory of Cape Parry, bearing E.N.E. from this place. The beach, on the south side of Harrowby Bay, not far from Point Maitland, was thickly strewed with fragments of dark red and of white sandstone, together with some blocks of the above-mentioned limestone, and a few boulders of sienite.

From Cape Bathurst the coast line has a S.E. direction, and is formed by precipitous cliffs, which gradually rise in height from thirty feet to six hundred. The beds composing these cliffs appear to be analogous to those of the alum-shale banks at Whitby, and similar to those which skirt the Scented-grass Hill and Great Bear Mountain, in Great Bear Lake. The Scented-grass Hill is distant from Cape Bathurst about three hundred miles, on a S.E. bearing, which corresponds, within a point, with the direction of the principal mountain chains in the country. There is evidently a striking similarity in the form of the ground plan of these two promontories. At the extremity of Cape Bathurst the cliffs consist of slaty-clay, which, when dry, has a light bluish-gray colour, a slightly greasy feel, and falls down in flakes. The rain-water had penetrated the cliff to the depth of three yards from the summit; and this portion was frozen, on the 17th July, into an icy wall, which crumbled down as it thawed. On proceeding a little further along the coast, some beds were observed that possessed, when newly exposed to the air, tenacity enough to be denominated stone, but which, under the action of water, speedily softened into a tenacious bluish-clay.

191

At Point Traill we were attracted by the variegated colours of the cliff, and on landing found that they proceeded from clays baked by the heat of a bed of bituminous-alum-
192 shale which had been on fire. Some parts of the earth
193 were still warm. The shale is of a brown colour and
197 thin slaty structure, with an earthy fracture. It con-
198 tains many interspersed crystals of selenite; between
199 its lamina there is much powdery alum, mixed with sulphur, and it is traversed by veins of brown selenite, in slender prismatic crystals. The bed was much broken down, and hid by the debris of the bank, but in parts it was several yards thick, and contained layers of the wax-
200 coloured variety of alum, named Rock-butter. The
194 shale is covered by a bed of stone, chiefly composed of
195 oval distinct concretions of a poor calcareous clay-iron
196 stone. These concretions have a straight cleavage in the direction of their short axis, and are often coasted by fibrous calc-sinter and calcedony. The upper part of the cliff is clay and sand passing into a loosely cohering sandstone. The strata are horizontal, except in the neighbourhood of ravines, or of consumed shale, when they are often highly inclined, apparently from partial subsidence. The debris of the cliff form declivities, having an inclination of from fifty to eighty degrees, and the burnt clays variously coloured, yellow, white, and deep red, give it much the appearance of the rubbish of a brick-field. The view of the interior, from the summit of the cliff, presents a surface slightly varied by eminences, which swell gently to the height of fifty or sixty feet above the general level. The soil is clayey, with a very scanty vegetation, and there are many small lakes in the country.

Ten miles further on, the alum-shale forms a cliff two hundred feet high, and presents layers of the Rock-butter about two inches thick, with many crystals of selenite on the surfaces of the slates. The summit of the cliff consists
201 of a bed of marly gravel two yards thick, which is composed of pebbles of granite, sienite, quartz, lydian-stone, and compact limestone, all coated by a white powdery marl. The dip of the strata at this place is slightly to the northward.

A few miles to the south-east of Wilmot Horton River the cliffs are six hundred feet high, and present acclivities having an inclination of from thirty to sixty degrees, formed of weathered slate-clay. Some beds of alum-shale are visible at the foot of these cliffs, containing much sulphate of alumina and masses of baked clay.

Two miles further along the coast the shaly strata were on fire, giving out smoke, and beyond this the cliffs become much broken but less precipitous, having fallen down in consequence of the consumption of the combustible strata. These ruined cliffs gradually terminated in green and sloping banks, whose summit was from one to two miles inland, and about six hundred feet above the sea level. Considerable tracts of level ground occurred occasionally betwixt these banks and the beach. Wherever the ground was cut by ravines, beds of slate-clay were exposed. On reaching the bottom of Franklin Bay, we observed the higher grounds keeping an E.S.E. direction until lost to the view, becoming, however, somewhat peaked in the outline.

SEA COAST.—LIMESTONE.

Parry's Peninsula, where it joins the mainland, is very low, consisting mostly of gravel and sand, and is there greatly indented by shallow bays, but it gradually increases in height towards Cape Parry. The bays and inlets are separated from the sea by beaches composed of rolled pieces of compact limestone; and which, although they are in places only a few yards across, are several miles in length. The northern part of Parry's Peninsula belongs entirely to a formation which appears from the mineralogical characters of the stone composing the great mass of the strata, and the organic remains observed in it, to be identical with the limestone formations of Lakes Winipeg and Huron.

On the north side of Sellwood Bay, in lat. 69° 42′,
202 cliffs about twenty feet high are composed of a fine-
204 grained* brownish dolomite, in angular distinct concretions, and containing corallines and veins of calc-spar. In the same neighbourhood there is a bed of
203 grayish-black compact luculite with drusses of calc-spar, very similar to the limestone which occurs in

* Specimens from Sellwood Bay.
202 Fine-grained dark brownish-gray dolomite, with corallines filled with white calc-spar.
203 Lucullite grayish-black, compact, and without lustre.
204 Gray dolomite.
205 A rolled piece, evidently of the same rock with the preceding, containing the impression of a *cardium*.
206

highly inclined strata at the "Rock by the River Side," on the Mackenzie, and in horizontal strata in an island near that rock, where it forms angular concretions.

After passing Sellwood Bay, the north and east shores of Cape Parry, and the islands skirting them, present magnificent cliffs of limestone, which, from the weathering action of the waves of the sea, assume curious architectural forms. Many of the insulated rocks are perforated. Between the bold projecting cliffs of limestone there are narrow shelving beaches, formed of its debris, that afford access to the interior. The strata have generally a slight dip to the northward, and the most common Rock is a yellowish-gray dolomite which has a very compact structure, but presents some shining facets of disseminated calc-spar. This stone, which is not to be distinguished by its mineralogical characters from the prevailing limestone of Lake Winipeg, and at the passage of *La cloche* in Lake Huron, forms beds six or eight feet thick, and is frequently interstratified with a cellular limestone, approaching to chert in hardness, and exhibiting the characters of rauchwacke. In some parts, the rauchwacke is the predominating rock, and has its cells beautifully powdered with crystals of quartz or of calc-spar, and contains layers of chert of a milky colour. The chert has sometimes the appearance of calcedony, and is finely striped.

208, 209 The extremity of Cape Parry is a hill about seven hundred feet high, in which beds of brownish dolomite, impregnated with silica, are interstratified with a thin-slaty, gray limestone, having a compact structure.* The vegetation is very scanty, and there are some spots covered with fragments of dolomite, on which there is not the vestige even of a lichen. Many large boulders of greenstone were thrown upon the N.W. point of Cape Parry. The islands in Darnley Bay, between Capes Parry and Lyon, are composed of limestone.

* Specimens from the Promontory of Cape Parry, which rises into a hill, seven hundred feet high. Strata dipping lightly to the northward.
207 Yellowish-gray dolomite, imperfectly crystalline, being similar to the limestone of Lake Winipeg.
208 Brownish dolomite impregnated with silica.
209 Thin-slaty, gray limestone. Very common also in Lake Winipeg.
210, 211 Boulders of dolomite.
212
213 Brown dolomite, with drusy cavities and veins, lined by calc-spar.

SEA-COAST.—FORMATION OF SLATE-CLAY, SANDSTONE, AND LIMESTONE, WITH TRAP-ROCKS.

FROM Cape Lyon to Point Tinney, the rocks forming the coast-line are slate-clay, limestone, greenstone, sandstone, and calcareous puddingstone.

214 Near the extremity of Cape Lyon the *slate-clay* predominates, occurring in straight, thin, bluish-gray layers, which are interspersed with detached scales of mica. It sometimes forms thicker slates, that are impregnated with 215 iron, and occurs alone, or interstratified in thin beds with a reddish, small-grained limestone. The strata, in general, dip slightly to the N.E., and form gently-swelling grounds, which at the distance of about fifteen miles to the southward terminate in hills, named the Melville Range. These hills are apparently connected with those which skirt the coast to the westward of Parry's Peninsula, have rather a soft outline, and do not appear to attain an altitude of more than seven or eight hundred feet above the sea. Ridges of naked trap-rocks, which traverse the lower country betwixt the Melville hills and the extremity of the Cape, rise abruptly to the height of one hundred or one hundred and fifty feet, and have, in general, an E.N.E. direction. When these trap ridges reach the coast, they form precipices which frequently have a columnar structure, and the nearly horizontal strata of slate-clay are generally seen underlying the precipices. In many places the softer clay strata are worn considerably away, and the columns of greenstone hang over the beach. Columns of this description occur at the north-eastern extremity of the Cape, and the slate-clay is not altered at its point of contact with the greenstone. The soil in this neighbourhood is clayey, and some small streams have pretty lofty and steep clayey banks; the shaly strata appearing only at their base. A better sward of grasses and carices exists at Cape Lyon, than is usual on those shores. Many boulders of greenstone and large fragments of red sandstone strew the beach.

At Point Pearce, four or five miles to the eastward of Cape Lyon, a reddish, small-grained limestone forms perpendicular cliffs two hundred feet high, in which a remarkable cavern occurs. Near these cliffs the slate-clay and reddish limestone are interstratified, and form a bold rocky point, in which the strata dip to the N.E. at an angle of 20°. The coast line becomes lower to the eastward, and at Point Keats a fine-grained, flesh-coloured sandstone occurs. This sandstone is quartzose, does not possess much tenacity, and is without any apparent basis.

At Point Deas Thompson the limestone re-appears, having reddish-brown and flesh-red colours, and a splintery fracture. There are some beautiful Gothic arches formed in the cliffs there by the weathering of the strata.

Five miles farther along the coast, near Roscoe River, the same kind of limestone forms cliffs twenty-five feet high, and is covered by thin layers of soft slate-clay. On the top of these cliffs we observed a considerable quantity of drift-timber and some hummocks of gravel. The spring tides do not rise above two feet. The Melville Range approaches within three miles of the coast there, and presents a few short conical summits, although the hills composing it are mostly round-backed.

At Point De Witt Clinton, a compact blackish-blue limestone, traversed by veins of calc-spar, forms a bed thirty feet thick, which reposes on thin layers of a soft, compact,
217 light, bluish-gray limestone or marl. The cliffs at this
218 place are altogether about seventy feet high, but their
219 bases were concealed by accumulations of ice. Veins filled with compact and fibrous gypsum traverse the upper limestone. Naked and barren ridges of greenstone, much iron-shot, cross the country here, in the same manner as at Cape Lyon. The soil consists of gravel and clay; the former mostly composed of whitish magnesian limestone; and the vegetation is very scanty.

At Point Tinney, in lat. 69° 20′, cliffs of a calcareous puddingstone, about forty feet high, extend for a mile along the coast. The basis, in most of the beds, is calc-spar; but in some small layers it is calcareous sand. The imbedded pebbles are smooth, vary in magnitude, from the size of a pea to that of a man's hand, and are mostly or entirely of chert, which approaches to calcedony, and, when striped, to agate in its characters. Perhaps, much of the gravel which covers the country is derived from the destruction of this conglomerate rock.

SEA COAST.—LIMESTONE.

From Point Clifton to Cape Hearne, the whole coast consists of a formation of limestone precisely similar to that which occurs on Lake Winipeg and Parry's Peninsula.

Dolomite, the prevailing rock in this formation, is generally in thin layers, and has a light smoke-gray colour, varying occasionally to yellowish gray, and buff. Its structure is compact, with little lustre, except from facets of disseminated calc-spar. It sometimes passes into milk-white chert, which forms beds. In some places the dolomite alternates with cellular limestone, which is generally much impregnated with quartz, and has its cavities powdered with crystals of that mineral. No organic remains were observed in the strata, but fragments, evidently derived from some beds of the formation, contained othoceratites, like those of Lake Huron. The strata, though nearly horizontal, appear to crop out towards the north and east, forming precipices about ten feet high, facing in that direction, and running like a wall across the country. In many places, however, and particularly at Cape Krusenstern, the strata terminate in magnificent cliffs upwards of two hundred feet high, the country in the interior remaining level. Mount Barrow is a small hill of limestone, of a remarkable form, being a natural fortification surrounded by a moat. The coast line is indented by shallow bays, and skirted by rocks and islands.

In the whole country occupied by this formation, the ground is covered with slaty fragments, sometimes to the depth of three feet or more. These slates appear to have been detached from the strata they cover, by the freezing of the water, which insinuates itself betwixt their layers. At Cape Bexley, the fragments of dolomite cover the ground to the exclusion of all other soil; and in a walk of several miles, I did not see the vestige of a vegetable, except a small green scum upon some stones that formed the lining of a pond which had dried up. In this neighbourhood there are a number of straight furrows a foot deep, as if a plough had been drawn through the loose fragments. After many conjectures as to the cause of this phenomenon, I ascertained that the furrows had their origin in fissures of the strata lying underneath.

At the commencement of this formation between Point Tinney and Point Clifton, the coast is low, and a stream of considerable magnitude, named Croker River, together with many rivulets, flow into the sea. Its termination to the southward of Cape Hearne is also marked by a low coast line, which is bounded by the bold rocky hills of Cape Kendall.

FORMATION SIMILAR TO THAT AT CAPE LYON.

The beach between Cape Hearne and Cape Kendall is in some places composed of slate-clay, and of a clay resembling wacke. Many large boulders of greenstone occur there. Cape Kendall is a projecting rocky point, about five or six hundred feet high, and nearly precipitous on three sides, which are washed by the sea. On the north, its rocks consist entirely of greenstone, but on the south side of the Cape the greenstone in lofty columns reposes on thin-slaty beds of fine-grained, bluish-gray limestone. Back's Inlet presents on each side a succession of lofty precipitous headlands, which have the shape termed, by seamen, "the gunner's quoin." Most of the islands and points near the mouth of the Coppermine have this form, and are composed of trap rocks. One of Cowper's 220 islands on which we landed consists of beds of greenstone cropping out like the steps of a stair.

A low ridge of greenstone exists at the mouth of the Coppermine river, and from thence to Bloody-fall, a distance of ten miles, the country is nearly level, with the exception of some low ridges of trap which run through it. The channel of the river is sunk about one hundred and fifty feet below the surrounding country, and is bounded by cliffs of yellowish white sand, and sometimes of clay, from beneath which, beds of greenstone occasionally crop out.

At Bloody-fall, a round-backed ridge of land, seven or eight hundred feet high, crosses the country. It has a gentle ascent on the north, but is steep towards the south. The river at the fall makes its way through a narrow gap, whose nearly precipitous sides consist of tenacious clay, the bed and immediate borders of the stream being formed of greenstone.* From thence to the Copper Mountains, gently undulated plains occur, intersected in various parts by precipitous ridges of trap rocks, and the river flows in a narrow chasm, sunk about one hundred feet below their level. A few miles above Bloody-fall, strata of light gray clay-slate, dipping to the north-east, at an angle of 20°, support some greenstone cliffs on the banks

* In the geological notices appended to the narrative of Captain Franklin's Journey to the Coppermine, I have termed this rock a dark purplish-red felspar rock. On examining it again on this journey, I perceived it to be a greenstone, whose surfaces weather of a rusty brown colour.

of the river. From this place to the Copper Mountains
222 the rocks observed in the ravines were a dark reddish-
223 brown, felspathose sandstone, and gray slate-clay in ho-
224 rizontal strata, with greenstone rising in ridges. The
soil is sandy, and in many places clayey, with a pretty close grassy sward. Straggling spruce trees begin to skirt the banks of the river about eighteen or twenty miles from the sea.

COPPER MOUNTAINS.

The Copper Mountains rise perhaps eight or nine hundred feet above the bed of the river, and at a distance, present a somewhat soft outline, but on a nearer view they appear to be composed of ridges which have a direction from W.N.W. to E.S.E. Many of the ridges have precipitous sides, and their summits, which are uneven and stony, do not rise more than two hundred, or two hundred and fifty feet above the vallies, which are generally swampy and full of small lakes. The only rocks noticed when we crossed these hills on the late journey, were clay-slate, greenstone, and dark red sandstone, sometimes containing white calcareous concretions, resembling an amygdaloidal rock. On our first journey down the Coppermine River, we visited a valley where the Indians had been accustomed to look for native copper, and we found there many loose fragments of a trap rock, containing native copper, green malachite, copper glance, and iron-shot copper green; also trap containing greenish-gray prehnite with disseminated native copper, which, in some specimens was crystallized in rhomboidal dodecahedrons. Tabular fragments of prehnite, associated with calc-spar and native copper, were also picked up, evidently portions of a vein, but we did not discover the vein in its original repository. The trap-rock, whose fragments strewed the valley, consists of felspar, deeply coloured by hornblende. A few clumps of white spruce trees occur in the vallies of the Copper Mountains, but the country is in general naked. The Coppermine River makes a remarkable bend round the end of these hills.

After quitting the Copper Mountains, and passing a valley occupied by a chain of small lakes in lat. 67° 10′, long. 116° 45′, we travelled over a formation whose prevailing rocks are spotted sandstone and conglomerate, and which forms the *height of land* betwixt Bear Lake and the Coppermine River. The ascent to this height from the eastward is gradual, but the descent towards Bear Lake is more rapid. The country is broken and hilly, though the height of the hills above the sea is perhaps inferior to that of the Copper Mountains. The vallies through which the small streams that water the country flow, are narrow and deep, resembling ravines, and their sides are clayey. The ground is strewed with gravel.

The *sandstone* has very generally a purplish colour, with gray spots of various magnitudes. It is fine-grained, hard, has a somewhat vitreous lustre, and contains little or no disseminated mica.

The *conglomerate* consists of oval pebbles of white quartz, sometimes of very considerable magnitude, imbedded in an iron-shot cement. Many of the pebbles appear as if they had been broken and firmly re-united again. The conglomerate passes into a coarse sandstone.

Porphyry and granite form hills amongst the sandstone strata.

The *porphyry* has a compact basis, like hornstone, of a dull brown colour, which contains imbedded crystals of felspar and quartz, and occasionally of augite. It forms some dome-shaped and short conical hills.

The *granite* is disposed in oblong ridges, with small mural precipices. It has, generally, a flesh-red colour, and contains some specks of augite, but little or no mica. The granite and porphyry were observed only on the east side of the height of land, the brow of which, and its whole western declivity, is formed of sandstone. Boulders of granite and porphyry, precisely similar to the varieties which occur *in situ* on the height of land, are common on the beach at Fort Franklin, and on the banks of the Mackenzie above Bear Lake.

To the westward of the height of land, the country on the banks of Dease River is more level, and few rocks *in situ* were seen, until within five or six miles of Bear Lake, where the stream flows through a chasm, whose sides are composed of a soft, fine-grained red sandstone, like that which occurs in the vale of Dumfries, in Scotland. Several ravines here have their sides composed of fine sand, inclosing fragments of soft sandstone.

About three miles from the mouth of Dease River we came to a limestone formation, which has been already noticed in the account of the geological structure of the shores of Great Bear Lake.

EASTERN CHAIN OF PRIMITIVE ROCKS.

The preceding part of the paper describing the rock formations which were noticed on the route of the expedition from Great Slave Lake down the Mackenzie along the shores of the Arctic Sea, the Coppermine, Great Bear Lake, and Great Bear River, being a distance of three thousand miles, I shall, by way of supplement, mention very briefly some of the more southern deposits.

The first I have to speak of is the chain of primitive rocks to which I have alluded in page 289, as extending for a very great distance in a north-west direction, and inclining in the northern parts slightly towards the Rocky Mountain Chain. Dr. Bigsby, in his account of the geology of Lake Huron says, that " The primitive rocks on the northern shores of that lake are part of a vast chain, of which the southern portion, extending probably uninterruptedly from the north and east of Lake Winipeg, passes thence along the northern shores of Lakes Superior, Huron, and Simcoe, and after forming the granitic barrier of the Thousand Isles, at the outlet of Lake Ontario, spreads itself largely throughout the state of New York, and there joins with the Alleghanies, and their southern continuations." It is not my intention to say any thing further of the rocks in the districts of which Dr. Bigsby speaks, although in travelling from the United States to Lake Winipeg the expedition passed over them. That zealous geologist has already given, in various publications, many interesting and accurate details of the formations on the borders of the great lakes; an account of those which lie some degrees farther to the north is inserted in the second volume of the Geological Transactions,—and there are some notices of them in the Appendix to the narrative of Captain Franklin's First Journey. My object at present is, merely to trace the western boundary of the primitive rocks in their course through the more northerly parts of the American continent.

I have already quoted Sir Alexander Mackenzie's original and important remark, of the principal lakes in those quarters being interposed betwixt the primitive rocks and the secondary strata, lying to the westward of them—Lake Winipeg is an instance in point. It is a long, narrow lake, and is bounded throughout on its east side by primitive rocks, mostly granitic, whilst its more indented western shore is formed of horizontal limestone strata. The western boundary of the primitive rocks, extending on this lake about two hundred and eighty miles, has nearly a north-north-west direction. From Norway Point, at the north end of the lake, to Isle à la Crosse, a distance of four hundred and twenty miles in a straight line, the boundary has a west-north-west direction. For two hundred and forty miles from Isle à la Crosse to Athabasca Lake, the course of the primitive rocks is unknown to me; but from Athabasca Lake to M'Tavish's Bay, in Great Bear Lake, a distance of five hundred miles, their western edge runs about north-west-by-west, and is marked by the Slave River, a deep inlet on the north side of Great Slave Lake, and a chain of rivers and lakes, (including great Marten Lake,) which discharge themselves into that inlet.

Captain Franklin on his voyage crossed this primitive chain nearly at right angles to its line of direction, in proceeding from Hudson's Bay to Lake Winipeg—it was there two hundred and twenty miles wide.

The hills composing the chain are of small elevation, none of them rising much above the surrounding country. They have mostly rounded summits, and they do not form continuous ridges; but are detached from each other by vallies of various breadth, though generally narrow, and very seldom level. The sides of the hills are steep, often precipitous. When the vallies are of considerable extent, they are almost invariably occupied by a lake, the proportion of water in this primitive district being very great; from the top of the highest hill on the Hill River, which has not a greater altitude than six hundred feet, thirty-six lakes are said to be visible. The small elevation of the chain may be inferred from an examination of the map, which shows that it is crossed by several rivers, that rise in the Rocky Mountains, the most considerable of which are the Churchill and the Saskatchewan, or Nelson River. These great streams have, for many hundred miles from their origin, the ordinary appearance of rivers, in being bounded by continuous parallel banks; but on entering the primitive district, they present chains of lake-like dilatations, which are full of islands, and have a very irregular outline. Many of the numerous arms of these expansions wind for miles through the neighbouring country, and the whole district bears a striking resemblance, in the manner in which it is intersected by water, to the coast of Norway and the adjoining part of Sweden. The successive dilatations of the rivers have scarcely any current, but are connected to each other by one or more straits, in which the water-course is more or less obstructed by rocks, and the stream is very turbulent and rapid. The most prevalent rock in the chain is gneiss; but there is also granite and mica-slate, together with numerous beds of amphibolic rocks.

LIMESTONE OF LAKE WINIPEG.

To the westward of the chain of primitive rocks, through a great part, if not through the whole of its course, lies an extensive horizontal deposit of limestone.

Dr. Bigsby, in the Geological Transactions, has described, in detail, the limestone of Lake Huron, and is disposed to refer "the cavernous and brecciated limestone of Michilimackinac to the magnesian breccia, which is in England connected with the red marl;" whilst the limestones of St. Joseph, and the northern isles, he considers as more resembling the well-known formation of Dudley, in Staffordshire. The limestone of Thessalon Isle, in which there occurs the remarkable species of orthoceratite which he has figured, he describes as decidedly magnesian. I observed this orthoceratite in the limestone strata of one of the isles forming the passage of La Cloche in Lake Huron. The limestone deposits of Lake Winipeg and Cape Parry exactly resemble that of La Cloche in mineralogical characters, and in containing the same orthoceratite which was also found by Captains Parry and Lyon at Igloolik.

The colour of the limestone of Lake Winipeg is very generally yellowish-white, passing into buff, on the one hand, and into ash-gray on the other. A reddish tinge is also occasionally observed. Much of it has a flat fracture, with little or no lustre, and a fine-grained arenacious structure. A great portion of it, however, is compact, and has a flat conchoidal and slightly splintery fracture. This variety passes into a beautiful china-like chert. Many of the beds are full of long, narrow vesicular cavities, which are lined sometimes with calc-spar, but more frequently with minute crystals of quartz.

1001, 1014

The beds of this formation seldom exceed a foot in thickness, and are often very thin and slaty. The arenacious and cherty varieties frequently occur in the same bed; sometimes they form distinct beds. The softer kinds weather readily into a white marl, which is used by the residents to whitewash their houses. Wherever extensive surfaces of the strata were exposed, as in the channels of rivers, they were observed to be traversed by rents crossing each other at various angles. The larger rents, which were sometimes two yards or more in width, were, however, generally parallel to each other for a considerable distance.

Professor Jameson enumerates *terebratulæ*, *orthoceratites*, *encrinites*, *caryophyllitæ*, and *lingulæ*, as the organic remains in the specimens brought home by Captain Franklin on his first expedition. Mr. Stokes and Mr. James De Carle Sowerby have examined those which we procured on the last expedition, and found amongst them *terebratulites*, *spirifers*, *maclurites*, and *corallines*. The maclurites belonging to the same species, with specimens from Lakes Erie and Huron, and also from Igloolik, are perhaps referrible to the *Maclurea magna* of Le Sueur. Mr. Sowerby determined a shell, occurring in great abundance in the strata at Cumberland-house, about
1015
1019
one hundred and twenty miles to the westward of Lake Winipeg, to be the *Pentamerus Aylesfordii.*

The extent to the westward of the limestone deposit of Lake Winipeg is not well known to me; but I have traced it as far up the Saskatchewan as Carlton House, and its breadth there is at least two hundred and eighty miles. For about one hundred miles below Carlton House, the river Saskatchewan flows betwixt banks from one to two hundred feet in height, consisting of clay or sand, and the beds of limestone are exposed in very few places. The plains in the neighbourhood of Carlton abound in small lakes, some of which are salt. The country which the Saskatchewan waters for one hundred and ninety miles before it enters Lake Winipeg, is of a different kind. It is still more flat than that about Carlton, and is so little raised above the level of the river, that in the spring-floods the whole is inundated, and in several places the river sends off branches which reunite with it after a course of many miles. In this quarter the soil is generally thin, and the limestone strata are almost every where extensively exposed. To the southward of Cumberland House, the Basquiau Hill has considerable elevation. I had not an opportunity of visiting it; but in the flat limestone strata, near its foot, there are salt springs, from which the Indians sometimes procure a considerable quantity of salt by boiling; and there are several sulphureous springs within the formation.

I observed no beds of conglomerate in it, and no sandstone associated with it; but the extensive plains which lie betwixt

Carlton House and the Rocky Mountains are sandy, and beds of sandstone are said to be visible in some of the ravines.

The line of contact of the limestone with the primitive rocks of Lake Winipeg, is covered with water; but at the Dog's-Head, and near the north end of Beaver Lake, they are exposed within less than a mile of each other. To the southward of the Dog's-Head in Lake Winipeg, and in a few other quarters, some schistose rocks, belonging to the transition series, are interposed between the two formations.

Before quitting the formations of Lake Winipeg, I may remark, that the height of that lake above the sea is perhaps equal to that of Lake Superior, which is eight hundred feet.

LIMESTONE OF THE ELK AND SLAVE RIVERS.

The next formation I have to mention is one which appears to possess most of the characters ascribed by German geologists to the zechstein. It extends from the north side of the Methy carrying-place down the Clearwater, Elk, and Slave Rivers, and along the south shore of Great Slave Lake to the efflux of the Mackenzie. The line I have traced was the route of the expedition, and is also very nearly that of the eastern boundary of the limestone. Primitive rocks occur in Lake Mammawee, Athabasca Lake, and on the Stony River; and on several parts of the Slave River they are separated from the limestone only by the breadth of the stream. On Great Slave Lake, the Stony Island, on the north-east side of the mouth of Slave River, is composed of granite, whilst the limestone strata are exposed at Fort Resolution on the south-west side.

The limestone in this extensive tract is commonly in thin and nearly horizontal beds, and much of it exactly resembles in mineralogical characters the dolomite and chert of Lake Winipeg. It is interstratified with thin beds of soft white marl; and in a few places with a marly sandstone. Extensive beds of stinkstone also occur, and many beds of limestone containing fluid bitumen in cavities. The bitumen is in such quantity, in some quarters, as to flow in streams from fissures in the rock; and in an extensive district, around Pierre au Calumet on the Elk River, slaggy mineral pitch fills the crevices in the soil, and may be collected in large quantities by digging a well.

1027, 1028

A calcareous breccia also exists in various places, particularly on the Slave River. Springs depositing from their waters sulphur, and sulphate of lime, slightly mixed with sulphate of magnesia, muriate of soda, and iron, are common and copious. A few miles to the westward of the Slave River, there is a ridge of hills several miles long, and about two hundred feet high, having several beds of compact, grayish gypsum exposed on its sides. From the base of this hill there issue seven or eight very copious, and many smaller springs, whose waters deposit a great quantity of very fine muriate of soda by spontaneous evaporation. The collected rivulets from these springs form a stream which is, at its junction with the Slave River, sixty yards wide and eight or ten feet deep.

1020 to 1026

1029 to 1032

The organic remains in this deposit, according a list kindly furnished by Mr. Sowerby, consist of *spirifers*, one of which is the *spirifer acuta;* several new *terebratulæ*, of which one resembles the *T. resupinata*, a *cirrus*, some crinoidal remains, and corals.

At the union of Clearwater and Elk Rivers, the limestone beds are covered to the depth of one hundred and fifty feet with bituminous shale.

I have stated, that on Slave River this limestone formation succeeds immediately to primitive rocks, but I am not acquainted with the rocks that lie to the eastward of it on the Elk River. The traders report that there are extensive deposits of sandstone on the eastern arm of the Athabasca Lake, and, perhaps, these sandstones extend nearly to Clearwater River. Sand covers the limestone on that river to the depth of eight or nine hundred feet, and the fragments of sandstone in it are large, numerous, and not worn.

The quantity of gypsum in immediate connection with extremely copious and rich salt springs, and the great abundance of petroleum in this formation, together with the arenacious, soft, marly, and brecciated beds interstratified with the dolomite, and above all, the circumstance of the latter being by far the most common and extensive rock in the deposit, led me to think that the limestone of the Elk and Slave Rivers was equivalent to the zechstein of the continental geologists. My opinion, however, on this subject is, from a total want of practical acquaintance with the European rock formations, of little weight; and several eminent geologists are, after an examination of the organic remains and mineralogical characters of the specimens brought home, inclined to consider the formation as analogous to the carboniferous or mountain-limestone of England.

As to the limestone formation of Lake Winipeg, I have no doubt of its identity with that occurring in the islands at the passage of La Cloche, in Lake Huron, and also with that at Cape Parry and at Cape Krusenstern, on the coast of the Arctic Sea. It is probable, also, that these four deposits belong to the same epoch with the limestone of Elk and Slave Rivers, although they differ in containing little or no petroleum. It is proper to mention, however adverse it may be to the opinion I have ventured to hint at above, of these extensive horizontal deposits of limestone being referable to the zechstein, that the limestone of Lake Huron is generally considered as belonging to the mountain-limestone; and Professor Jameson, from a review of the organic remains occurring in the Lake Winipeg deposit, considered that it also belonged to that formation. The formation of Cape Lyon may be, with less danger of a mistake, referred to the transition or mountain-limestone.

THE END.

IX.

AMERICAN JOURNAL OF THE MEDICAL SCIENCES,

PUBLISHED QUARTERLY

ON THE FIRST OF NOVEMBER, FEBRUARY, MAY, AND AUGUST.

Each Number will contain about two hundred and forty pages.

Terms five dollars per annum.

Gentlemen desirous to be supplied with it, are requested to transmit the amount of one year's subscription to the publishers, or to any of the agents. The demand for this work has so far exceeded the expectations of the publishers, as to render it necessary to reprint the first number.—Four Nos. have already appeared.

X.

NARRATIVE OF A JOURNEY

Through the Upper Provinces of India, from Calcutta to Bombay. By the Late Reginald Heber, D. D. Lord Bishop of Calcutta. In 2 vols. 8vo. with a map.

"It forms a monument of talent, sufficient, singly and alone, to establish its author in a very high rank of English Literature. It is one of the most delightful books in the language, and will, we cannot doubt, command popularity as extensive and as lasting as any book of travels that has been printed in our time. Certainly, no work of its class that has appeared since Dr. Clarke's can be compared to it for variety of interesting matter, still less for elegance of execution. The style throughout easy, graceful, and nervous, carries with it a charm of freshness and originality not surpassed in any personal memoir with which we are acquainted."—*Quarterly Review, No. 73.*

XI.

PALESTINE, AND OTHER POEMS,

By Bishop Heber. With a Memoir of his Life, 18mo.

"Bishop Heber's Palestine, and other Poems, place that lamented scholar in the first class of contemporary bards. The biographical sketch prefixed to the compilation, which is original, as such, does credit to the talent and feeling of the author."—*National Gazette.*

XII.

THE ROMANCE OF HISTORY.—ENGLAND.

By Henry Neele. In 2 vols. 12mo.

"Truth is strange—stranger than fiction."—*Byron.*

"The plan of this work is excellent. It consists of a tale, founded either on legendary lore, tradition, or historical fact, for every monarch's reign, from William the Conqueror to Charles the First, inclusive. It necessarily follows that there is great variety both of interest and character. The early monkish superstitions are succeeded by stern chivalry; and chivalry yields in turn to the gradual alteration of national manners, as we descend the stream of time to the latest period. Mr. Neele has bestowed great pains upon his many topics, and displays much ability in his treatment of them."—*Literary Gazette.*

"Mr. Neele has produced tales of the most vivid and intense interest."—*Literary Magnet.*

"His work is a valuable addition to all the Histories of England extant."—*Arliss' Magazine.*

"We recommend Mr. Neele's very interesting and clever work to the careful perusal of our readers." —*London Weekly Review.*

XIII.

PRIVATE MEMOIRS

Of the COURT of NAPOLEON. By L. F. J. De Bausset, former Prefect of the Imperial Palace. Translated from the French.

"The anecdotes of the imperial court are very garrulous and amusing."—*New Monthly Magazine.*

"There never was a more vigilant Paul Pry near a throne."—*Am. Quarterly Review.*

XIV.

THE FRENCH COOK.

By Louis Eustache Ude, Ci-devant Cook to Louis XVIII. and the Earl of Sefton, and Steward to his late Royal Highness, the Duke of York. First American, from the Eighth London edition.—In one vol. 12mo. With plates.

"Mr. Ude is, beyond competition, the most learned of cooks—even of French cooks."—*Lit. Gaz.*

XV.

LORD BYRON and some of his CONTEMPORARIES.

By Leigh Hunt. In 8vo.

"Mr. Leigh Hunt, however, is not one of these dishonest chroniclers. His position with regard to Lord Byron, and the long and intimate habits of intercourse with him which he enjoyed, enabled him to contemplate the noble poet's character, in all its darkness and brightness. Gifted, too, like the subject of his memoir, with very remarkable talents, he is much more to be relied on, both in his choice of points of view, and his manner of handling his subject. He is not likely to spoil a bon-mot, an epi-

XVI.

SAYINGS AND DOINGS, OR SKETCHES FROM LIFE.

Third series.

"Cousin William is, perhaps, the highest effort of the author's deservedly popular pen; and no greater encomium need be bestowed."—*Literary Gazette.*

XVII.

HUMOURS OF EUTOPIA;

A Tale of Colonial Times, by AN EUTOPIAN. 2 vols. 12mo.

XVIII.

ELIA.

Essays that have appeared under that signature in the London Magazine. First Series, Second Edition.

Contents.—The South Sea House.—Oxford in the Vacation.—Christ's Hospital five and thirty years ago.—The two races of Men.—New Year's Eve.—Mrs. Battle's opinion on Whist.—A Chapter on Ears.—All Fools' Day.—A Quaker Meeting.—The old and new Schoolmaster.—Valentine's Day.—Imperfect Sympathies.—Witches and other Night Fears.—My Relations.—Mackery End, in Hertfordshire.—Modern Gallantry.—The Old Benches of the Middle Temple. —Grace before Meat.—My first play.—Dream-Children; a reverie.—Distant Correspondents.—The praise of Chimney-Sweepers.—A complaint of the Decay of Beggars in the Metropolis.—A dissertation upon Roast Pig.—A Bachelor's complaint of the Behaviour of Married people.—On some of the old Actors.—On the Artificial Comedy of the last century.—On the Acting of Munden.

ELIA.—*Second series.*

Contents.—To Elia.—Rejoicings upon the New Year's coming of Age.—Reflections on the Pillory.—Twelfth Night, or What You Will.—The old Margate Hoy.—On the inconveniences resulting from being Hanged.—Letter to an Old Gentleman whose education has been neglected.—Old China.—On Burial Societies, and the Character of an Undertaker.—Barbara S——.—Guy Faux.—Poor Relations.—A Vision of Horns.—On the Danger of Confounding Moral with Personal Deformity.—On the Melancholy of Tailors.—The Nuns and Ale of Caverswell.—Valentine's Day.—The Child Angel.—Amicus Redivivus.—Blakesmoor in H———shire.—Detached Thoughts on Books and Reading.—Captain Jackson.—Confessions of a Drunkard.—The Old Actors.—The Gentle Giantess.—A Character of the late Elia.

XIX.

WASHINGTON IRVING'S WORKS.

KNICKERBOCKER'S NEW YORK. 2 vols. 12mo.
SKETCH BOOK. 2 vols. 12mo.
BRACEBRIDGE HALL. 2 vols. 12mo.
TALES OF A TRAVELLER. 2 vols. 12mo.

XX.

THE RED ROVER,

By the Author of the SPY, &c.

Also, a very handsome uniform Edition of the SPY, PIONEERS, PILOT, LIONEL LINCOLN, LAST OF THE MOHICANS, and PRAIRIE. In 12 vols. royal, 18mo.

XXI.

MR. BROUGHAM'S SPEECH

On the PRESENT STATE of the LAW. Corrected by himself. 8vo.

XXII.

WHIMS AND ODDITIES, IN PROSE AND VERSE,

With forty original designs, by THOMAS HOOD.

"But it is in vain to describe its companions; since we may leave it to the eye, which will discern more of the merits of these droll fancies at one glance, than we can convey any idea of in language, were we to fill the gazette."—*Literary Gazette.*

"Here we conclude; most heartily thanking Mr. Hood for the large quantum of ennui he has driven

CPSIA information can be obtained at www.ICGtesting.com
Printed in the USA
244112LV00005B/148/P

9 781241 692094